T0234695

SpringerBriefs in Petroleum Geoscience & Engineering

The SpringerBriefs series in Petroleum Geoscience & Engineering promotes and expedites the dissemination of substantive new research results, state-of-the-art subject reviews and tutorial overviews in the field of petroleum exploration, petroleum engineering and production technology. The subject focus is on upstream exploration and production, subsurface geoscience and engineering. These concise summaries (50-125 pages) will include cutting-edge research, analytical methods, advanced modelling techniques and practical applications. Coverage will extend to all theoretical and applied aspects of the field, including traditional drilling, shale-gas fracking, deepwater sedimentology, seismic exploration, pore-flow modelling and petroleum economics. Topics include but are not limited to:

- Petroleum Geology & Geophysics
- Exploration: Conventional and Unconventional
- Seismic Interpretation
- Formation Evaluation (well logging)
- Drilling and Completion
- Hydraulic Fracturing
- Geomechanics
- Reservoir Simulation and Modelling
- Flow in Porous Media: from nano- to field-scale
- Reservoir Engineering
- Production Engineering
- Well Engineering; Design, Decommissioning and Abandonment
- Petroleum Systems; Instrumentation and Control
- Flow Assurance, Mineral Scale & Hydrates
- Reservoir and Well Intervention
- Reservoir Stimulation
- Oilfield Chemistry
- Risk and Uncertainty
- Petroleum Economics and Energy Policy

Contributions to the series can be made by submitting a proposal to the responsible Springer contact, Anthony Doyle at anthony.doyle@springer.com, or the Academic Series Editor, Prof Dorrik Stow at dorrik.stow@pet.hw.ac.uk.

More information about this series at http://www.springer.com/series/15391

Bhajan Lal · Ali Qasim · Azmi Mohammad Shariff

Ionic Liquids in Flow Assurance

Springer

Bhajan Lal ⓘD
Chemical Engineering
Universiti Teknologi Petronas
Seri Iskandar, Perak, Malaysia

Ali Qasim ⓘD
Chemical Engineering
Universiti Teknologi Petronas
Seri Iskandar, Perak, Malaysia

Azmi Mohammad Shariff ⓘD
Chemical Engineering
Universiti Teknologi Petronas
Seri Iskandar, Perak, Malaysia

ISSN 2509-3126   ISSN 2509-3134   (electronic)
SpringerBriefs in Petroleum Geoscience & Engineering
ISBN 978-3-030-63755-2   ISBN 978-3-030-63753-8   (eBook)
https://doi.org/10.1007/978-3-030-63753-8

This Springer imprint is published by the registered company Springer Nature Switzerland AG
The registered company address is: Gewerbestrasse 11, 6330 Cham, Switzerland

Preface

Ionic liquids are eco-friendly chemicals having low melting point i.e. less than 100 °C, low vapor pressure, low toxicity, and high polarity. These are thermally and chemically stable as well. Ionic liquids have renownedly been known as molten salts at room temperature loosely bound by coordination of organic cation and anion together. The vast utilization of ionic liquids in almost all fields of chemistry and engineering is resulted to their above mentioned interesting properties which enable them as "green and sustainable chemicals" having tendency to dissolve wide range of inorganic and organic compounds. In oil and gas industry, ionic liquids have been employed in drilling operation, enhanced oil recovery (EOR), unconventional heavy oil recovery and flow assurance. Flow assurance is considered to be a critical task in oil and gas industry especially in the field of energy production in deep water. It is due to the harsh operational conditions of low temperatures and high pressures at long distances. Flow assurance industry has used different conventional chemicals to address the issue but the chemicals in current use are less effective at high pressure, temperature and in highly saline conditions. Ionic liquid, due to exceptional properties of negligible vapor pressure, non-flammability, and low melting point can prove to be very resourceful in midstream operations of flow assurance as well. They have shown a strong potential as substitutes for other compounds used in flow assurance applications. This book consists of four chapters discussing different aspects of ionic liquids applications in flow assurance in oil and gas industry.

The general application of ionic liquids and physiochemical properties with the focus on their application in flow assurance in oil and gas industry is discussed in the book. Furthermore, the role of ionic liquids as gas hydrate inhibitors in offshore pipelines to ensure safe flow has been taken into consideration. Gas hydrate occurrence poses a major threat to pipeline integrity. Different categories of gas hydrate inhibitors and the main factors influencing ionic liquids during gas hydrate inhibition are examined. The use of ionic liquids as corrosion inhibitors and their application in flow assurance industry to mitigate corrosion is considered as well. In this aspect, factors affecting corrosion inhibition performance of ionic liquids are discussed.

Lastly, ionic liquids application in wax, scale and asphaltenes deposition control is reviewed. The use of ionic liquids in this particular field is novel and has a potential to be explored further.

Perak, Malaysia

Bhajan Lal
Ali Qasim
Azmi Mohammad Shariff

Contents

Chapter 1
Ionic Liquids Usage in Oil and Gas Industry

1.1 Introduction to Ionic Liquids

Ionic liquids (ILs) are chemicals which are considered environmentally benign solvents that can be modified or tuned for specific requirements and applications. The physicochemical properties and the ability to tune into different combination of anions and cations are some of the unique characteristics of ILs which distinguish them from the other organic compounds [1, 2]. A physically conventional way to differentiate between ionic liquids and the similar 'molten salts' is through its melting point, i.e. ionic liquids are in liquid state at near ambient temperatures. As with any convention, there are exclusions, as there are molten salts that have even lower melting points than certain ILs. A more principle distinction between them is that ILs contain organic cations instead of inorganic cations.

The most salient property regarding ILs is the low melting point or liquidus temperature. Owing to the wide variety of ion combinations possible, ILs may possess melting points from as low as −90 °C. Examples of organic cation ILs include quaternary ammonium, phosphonium, pyridinium, and imidazolium salts [3–5]. The advantage of the organic salt's liquidus temperature has allowed it to be explored in combination with other anions for various applications [6]. Ionic liquids offer an advantage over other molecular solvents due to their low melting points and tailor-made size and shape for task-oriented applications. These properties render a wide range of utilization in various applications.

Ionic liquids have been employed in various scientific fields of research, more so in recent times [7, 8]. Their application includes in biomedical, pharmaceuticals, materials chemistry as catalysts, oil and shale processing, separation of petrochemical components, i.e. fractionation and as solvents in electrochemistry. The main advantage in using ILs in catalysis is that they are considered green catalysts. Figure 1.1 summarizes the scientific application of ionic liquids in numerous fields of study and their subfields.

B. Lal et al., *Ionic Liquids in Flow Assurance*,
SpringerBriefs in Petroleum Geoscience & Engineering,
https://doi.org/10.1007/978-3-030-63753-8_1

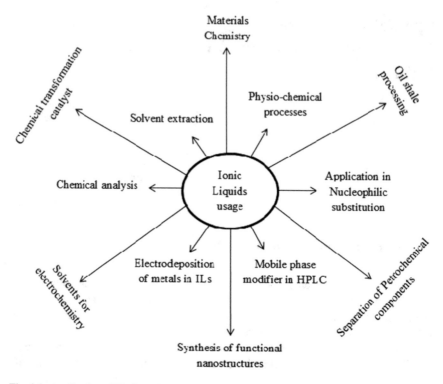

Fig. 1.1 Application of ILs in various scientific fields

Ionic liquids show interesting properties, such as a low melting point (conventionally determined as less than 100 °C), high viscosity, low vapour pressure, low toxicity, high polarity, good thermal stability and chemical stability. Such properties make these salts an effective alternative to corrosion inhibitors such as benzene and toluene, which show high volatility and toxicity [9]. Generally more viscous than other solvents, ILs show an increase in viscosity as its alkyl chain length increases. The melting point variation of ILs depends on both the cations and anions present. As the size of the anion or cation increases, the melting point decreases.

In the field of green chemistry, research and development towards processes that reduce the use of toxic materials is continually growing. While the field is only relatively emergent, researchers are focusing on environment-friendly techniques and processes. It includes the employment of eco-friendly compounds and chemicals. The main material of interest for research was in natural plant extracts and organic compounds [10]. Nevertheless, the process of extraction and purification of plants proved to be economically unfeasible and required large amounts of organic solvents and time. Consequently, studies have investigated the properties and synthesis process of cost-effective and non-toxic green chemicals [11]. In this regard, the emerging utilization of ionic liquids (ILs) is an important step towards addressing this issue. Ionic liquids provide a range of tailor-designed chemicals as there exist various varieties of cations and anions among them. The compounds with different combination of anions and cations can be developed for specific applications. Owing to these properties, it is advantageous to use them in catalysis and as solvent in mixtures [12]. Due to their high solubility in organic and inorganic mixtures, they are considered as sustainable, eco-friendly chemicals in the field of chemistry and chemical engineering [13].

1.2 Physicochemical Properties of Ionic Liquids

The simplest ionic liquids consist of a single cation and single anion. More complex examples can also be considered, by combining of greater numbers of cations and/or anions, or when complex anions are formed as the result of equilibrium processes. The structure of an ionic liquid plays a role in determining its melting point and liquidus ranges [1]. Within a case study involving dialkylimidazolium chloroaluminate, fundamentals explaining the effect of ionic liquid structure on melting points were studied [14]. Exploitation of the differences held by different ion combinations enables ionic liquids with a wide range of properties to be designed. The charge, size and distribution of charge on the respective ions are the main factors that influence the melting points of any generic salt; though small changes in the shape of uncharged, covalent regions of the ions can have an important influence on the melting points of the salts. The dominant force in ionic liquids is Coulombic attraction between ions.

1.2.1 Melting Point

The term liquidus temperature range is used alongside melting point in the description of ionic liquids. For example, water has a liquidus range of 100 °C (whereby it is liquid from 0 to 100 °C). Differential scanning calorimetry (DSC) is an effective way to measure transition temperatures, though X-ray scattering and polarizing microscopy have also proven to be efficient methods. Comparatively to other organic chemicals, ionic liquids exhibit complex thermal behaviour, and their melting points depend on the choice of cationic and anionic moieties [1].

1.2.2 Anion and Cation Properties

The choice of anionic and cationic parts of ionic liquids plays a significant role in defining their features. Anions and cations are dominated by Coulombic charge-charge attractions. Changes in the cations that do not help in stabilizing the Coulombic attractions of the crystal result in decreasing heat of formation and melting points of the chemical. With longer chain lengths, the presence of amphiphilic nature (containing both hydrophilic and hydrophobic behaviour) is manifested, showing layered lattices and hydrophobic van der Waals forces [8].

Regarding cation symmetry in ionic liquids, it is observed that melting points of symmetrically substituted cations is higher than the unsymmetrical cations, but it starts decreasing as alkyl group comprises of 8–10 carbons as it is considered to be a critical point. After this point, melting point further increases with increasing substitution. It is noticed that the melting point is dependent on alkyl substitution and ion symmetry of the ionic liquid.

The presence of several anions in these ionic liquids has the effect of significantly decreasing the melting point. It is observed due to the reason that covalency of ions increases and decrease in Coulombic attraction contribution to crystal lattice energy. As the size of anionic moiety increases in ionic liquids, melting point is reduced.

However, in terms of hydrogen-bonding ability, no clear pattern exists for its influence over the IL melting point. It is usually considered that the chemicals with lower melting points show inability to make hydrogen bond also. But in case of ionic liquids, many among them with lower melting points form hydrogen bonds just as those with high melting point which are longer chain. The effects of van der Waals interactions through the methyl group, or methyl-π interactions, etc., have a bigger effect than the electrostatic interactions through the alkyl-hydrogen bonds.

The main draw for using ionic liquids the various temperatures in which it can remain as a stable liquid, particularly within the harsher conditions where crude oils are located. It should be noted that the forces and interactions that govern the melting points of ionic liquids should not be looked at in isolation; as the way it interacts with other components within the medium such as will also control the dissolution and solubility of other components in the ionic liquids. Therefore, for its purpose as an additive, differing aqueous phase concentrations may have an effect on its melting point, and therefore its efficacy.

1.2.3 Viscosity of Ionic Liquids

The viscosity of a fluid is the representation of the internal friction of a fluid, and it largely influences the resistance that the fluid has against external surfaces or medium during flow. The two broad classifications of Newtonian and non-Newtonian fluid are mainly separated as such due to their differences in viscosity behaviour. Newtonian fluids such as water or air possess an almost unchanging viscosity regardless of strain rate, where its shear stress is directly proportional to the strain rate. Fluids which are non-Newtonian have apparent viscosities that depend on the shear rate and will stretch or thin against varying strain rates.

Ionic liquids are considered to be more viscous than most common molecular solvents due to its larger molecular size, though they are treated as Newtonian fluids. It is due to the reason that ILs density is independent to change in temperature and not affected significantly. At STP conditions, the viscosities of ILs usually vary within the range of 10 cP–500 cP. In comparison, the viscosities of mono ethylene glycol, glycerol, and water at 25 °C are found to be 16.1 cP, 934 cP and 0.89 cP respectively [15]. The ability of ionic liquids in altering the viscosity of raw oil is largely main mechanism of interest in studies, as it has the direct impact of improving the crude oil mobility, and thus its extraction.

1.3 Applications of ILs in Oil and Gas Industry

The application of ILs within the oil and gas industries (specifically for purposes apart from flow assurance) is mostly related to the extraction of naphthenic acids, removal of contaminants, desulfurization of fuels, denitrogenation of gasoline, and as a demulsification agent. ILs have also been employed in selective gas separation and mercury removal from natural gas (NG), biofuel synthesis or production, prospective usage in enhanced oil recovery (EOR), extraction of heavy oil or bitumen with ILs, use as wax and asphaltene inhibitors, CO_2 capturing and sequestration, applications of deep eutectic solvents (DES) in oil fields. Broadly considering, the use of ILs in the oil and gas industry can be categorized into four different sub-categories [13], as shown in Fig. 1.2.

1.3.1 Ionic Liquids in Drilling Operation

The standard drilling operations will require the use of drilling fluids to assist in the process. A typical risk associated with this operation is the leakage or loss of the fluid into the formation, which can cause instability in not only in the borehole itself, but also the drilling fluid properties [16]. To prevent this problem, additives controlling fluid loss along with drilling fluid are employed. It diminishes fluid leaking hazard.

In drilling, water-based fluids pose the issue of clay and shale hydration also. The use of ILs has the potential to inhibit the problem of shale hydration and swelling. A study by Yang et al. [17] for the ionic liquid of 1-vinyl-3-ethylimidazolium

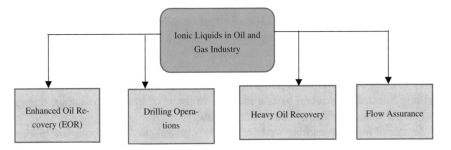

Fig. 1.2 Applications of ILs in offshore industry

bromide (VeiBr) monomer and its homopolymers has shown shale hydration inhibition behaviour for reducing instabilities in reservoirs for up to 300 °C [17]. Luo et al. [18] studied the effect of 1-octyl-3-methylimidazolium tetrafluoroborate (labelled as ILB) as shale hydration inhibitors for sodium montmorillonite drilling fluid. In hot-rolling dispersion test, it was found to possess excellent inhibitive properties when reaching a temperature of 160 °C. The shale inhibition performance of 0.05% ILB was also tested against two different samples, which are 5% KCl and 2% polyether diamine (PA). It was found that while the inhibition properties of 5% KCl performed the worst out of the three, the 0.05% ILB and 2% PA performed similarly. However, the lower concentration of ILB required to achieve the similar results showed by the conventionally popular PA could prove ILB as a more economically efficient option for use in shale hydrate inhibition. The ILB was also found to function well in ensuring minimal loss in the filtration property of the drilling fluid even at high temperatures.

Proceeding the studies above, Jia et al. [19] conducted an exploration of deep eutectic solvents (DESs) as close alternatives to ILs for the same purpose of shale inhibition. The three DESs studied were: CM-DES, CP-DES, and CIM-DES. CM-DES contains a 1:1 M ratio mixture of choline chloride and propane dioic acid, CP-DES comprises 2:1 M ratio of choline chloride and 3-phenylpropionic acid, and CIM-DES consists a molar ratio of 2:1:1 of choline chloride, itaconic acid, and 3-mercaptopropionic acid. The shale inhibition tests were conducted with sodium-bentonite (Na-bent) as the drilling fluid. It was seen that all three DESs at concentrations of 1% weight-per-volume (w/v) were able to outperform other reference fluids used (deionized water, 5 w/v% KCl, and 2 w/v% PA) in nearly all tests. In the shale inhibition tests, CP-DES showed the lowest apparent viscosity (56 mPa·s) and yield point (49 Pa), and the DESs results were also further confirmed by the inhibitive capacities through sedimentation volume measurement. Their strong overall performance is attributed to its ability to reduce the repulsion forces between the clay particles due to the strong electrostatic attraction between the choline cations in DES and clay, as well as the hydrogen bond formation between the functional groups within the DESs and hydroxyl or oxygen content on the clay surfaces. Even considering the cost, the low dosage required of DESs have shown that it can be effectively introduced as high-performance shale inhibitors due to just how quantitatively capable it is compared to conventional choices of KCl and PA.

In another study, Luo et al. [18] looked into another ionic liquid and its effects on the rheology and filtration properties of drilling fluids under high temperatures. The addition of the water-soluble IL improved both the rheological and filtration properties of the water-based mud (WBM) Na-montmorillonite at high temperatures of up to 240 °C. When investigating the addition of IL into the polymer water-based drilling fluids, it was found to also improve in terms of filtration volumes, though only at IL concentrations of below 0.05 mass%.

In case of high temperature wells, researchers investigated the rheological properties of an IL namely 1-butyl-3-methylimidazolium chloride (BMIM-Cl) in presence of WBM drilling fluid polymer [20]. The rheological properties of the IL-added WBM were found to improve, however only up to 180 °C. At 200 °C, the addition of

Table 1.1 ILs and their application in drilling operations [13]

Ionic liquids (ILs)	Application (inhibitor)	Advantage
1-Vinyl-3-ethylimidazolium bromide	Shale hydration	Drilling fluid thermal stability is improved up to 300 °C with enhanced shale hydration inhibition
1-Octyl-3-methylimidazolium tetraflfluoroborate	Shale hydration	Improved Na-montmorillonite inhibition properties
CM-DES, CP-DES, and CIM-DES	Shale hydration	As an additive, enhanced water-based drilling fluid inhibition abilities
1-Butyl-3-methylimidazolium chloride	Filtration loss	Drilling mud quality was improved in the pressure range of 1000 psi and temperature range of 25–200 °C
1-Octyl-3-methylimidazolium bromide	Shale inhibition	At lower concentrations, reduced surface tension and capillary action

BMIM-Cl into the WBM reduced its mud carrying capacity, indicating that caution must be taken in the selection of the anion and cation pairs for the IL to be used as additives. For other properties such as mud shear stress, viscosity, gel strength, and mud filtration behaviour; the addition of BMIM-Cl has generally positively contributed to their performance, across the tested temperature ranges of 20–200 °C.

From the cases looked into, ionic liquids have shown to be suitable substitutes as drilling fluid additives. Its addition to drilling fluids has generally contributed towards the reduction of shale viscosity, increased filtration, and higher mud carrying capacities. Certain ILs are also used to improve the effectiveness of drilling fluids when operating under higher temperature conditions. Many studies have also reported an economic advantage as the additional benefit to using ILs as additives. Furthermore, incompatible use of ILs may even result in a detriment to the performance of the drilling fluid. Table 1.1 highlights the types of IL surveyed. It should be noted however that the results established so far are recorded within small-scale, laboratory experimental settings. Within the results displayed across the studies, a larger test that has yet to take place is of field testing with the ILs, to document for its effectiveness in less-controlled environments, and any further unknown challenges that can affect the potential of future developed ILs for use in drilling operation.

1.3.2 Ionic Liquids in Enhanced Oil Recovery (EOR)

Within the oil and gas industry, EOR is employed as the tertiary type of oil recovery. The purpose of its implementation is in recovering additional amounts of oil that may still lie dormant in the well, yet is unobtainable by primary or secondary means. In terms of quantity, EOR is used to obtain around 30–60% of the remaining oils, and under favourable conditions may even recover up to 80% OIIP (oil initially in place) [1].

In EOR, at severe conditions of high salinity and temperature, the selectivity of a suitable chemical is one of the main challenges faced by the industry. As discussed by Bera and Belhaj [21], ILs are potentially superior compared to surfactants not only for its capabilities in reducing interfacial tension (IFT) and viscosity, but also its higher stability in wider temperature ranges and storage conditions, and higher tendency for environmental friendliness. Its use in the industry is largely held back by the lack of economically viable solutions and full field testing results.

Nandwani et al. [22] studied three variants of long-chain 1-alkyl-3-methyl imidazolium bromide (CnmimBr) for use as surfactants in EOR. Alkyl chain lengths of $n = 12, 14$ and 16 were investigated, and the best-performing IL was compared against the more conventional cationic surfactant, Cetyl trimethylammonium bromide (CTAB). The key parameter of critical micelle concentration (CMC) was used to identify the maximum point in which further increase in concentration of the surfactant would not show any more appreciable change in performance. The author's work agreed with the literature, as C16mimBr was reported to show the smallest CMC, by a significant margin compared to C12mimBr and C14mimBr [23, 24]. Thus, as alkyl chain length increases, its IFT reduction capabilities are stronger, while requiring a much smaller dosage. Compared to CTAB, C16mimBr showed a greater efficiency for oil recovery, and it is theorized to be due to its bulky heterocyclic ring (compared to CTAB's smaller quaternary ammonium cation), aromatic nature of the C16mimBr cations, leading to a stronger interaction between C16mimBr and the oil.

An important mechanism in EOR involves the alteration of wettability (e.g. from oil-wet to water-wet) and IFT between the reservoir solids and the injected fluids, to promote better mobility of the residual oils for recovery. Velusamy et al. [25] studied 6 imidazolium ionic liquids of incrementing alkyl chain lengths, and its influence over the wettability and IFT of alkane-water systems under different saline and environmental conditions for a quartz-alkane-aqueous IL system. At increasing temperatures, IFT has shown to decrease regardless of the alkyl length of the IL used, with an overall reduction of around 2.46–2.67 mN/m from its original value. More interestingly, it was observed that the addition of sodium chloride (NaCl) into the system showed further reduction in IFT, suggesting that ILs can be considered optimal alternatives compared to traditional surfactants in high salinity conditions. In the contact angle analysis for freshwater conditions, increasing the IL alkyl chain length showed an increase in contact angle, showing that the quartz surface wettability has shifted from oil-wet to water-wet to promote better oil recovery. The addition of saline conditions however contributed to a decrease in contact angle, with higher NaCl concentrations. The authors noted that the abundance of NaCl prevented the IL molecules from dissolving in the aqueous phase, thus blocking the adherence of alkanes on the oil surfaces.

Nabipour et al. [26] conducted a similar investigation of oil recovery properties within crude oil and aqueous saline conditions, using the ILs [C12mim] [Cl] and [C18mim] [Cl]. It was noted that [C12mim] [Cl] showed the best IFT and with a significantly lower CMC value of 50 ppm when tested in brine, compared to distilled water in which the IFT did not reach a critical concentration even when up to 4000 ppm of the IL was used. In the results of the flooding experiment, it was reported that [C12mim] [Cl] could achieve tertiary oil recoveries of up to 12.7%. Additional insight from the study reveals that the effect from wettability alteration and IFT on the ultimate oil recovery could be approximated to be about 36.7 and 63.2%, respectively. While this is very much specific to the IL in study, this led to the deduction that the IFT reduction was two times more significant than wettability alteration in improving the ultimate oil recovery within this case.

Abdullah et al. [27] studied four poly ionic liquids (PILs) for EOR applications in medium Arabian crude oil, with seawater as the aqueous phase of the dilution fluid. The prepared PILs have shown to have low thermal and salt sensitivity, and this was attributed to the formation of organic salts for the cations of the PILs, rather than the formation of polyelectrolytes [28]. In further tests, the wettability of the system showed gradual change towards a water-wet state as indicated from the decreasing contact angle and IFT, as the concentration of the tested PILs increased.

The main issue addressed by ILs for EOR is to widen the suitability of the surfactants for various conditions not typically by thermal recovery or miscible processes. Within the class of EOR using chemical additives (which ILs are typically placed under), some of the important factors to consider are the concentration requirement of chemicals, potentially adverse effect of saline conditions, and stability in thermal conditions. Of course, the key factor to consider is whether the IL could improve mobility control of the fluids at all, either through wettability alteration or IFT reduction. Table 1.2 lists several tested ionic liquids in EOR for different types of reservoir samples with the additional oil recovery being highlighted.

As mentioned in the review by Bera et al. [13], the demand for stable active chemicals within EOR is still present. While many studies have shown the prospect for newer chemical EOR techniques using IL, it still shares the similar sentiment of requiring further economic analysis and field testing before commercial utilization can be comfortably proceeded. It is undeniable however that the wettability altering and IFT reducing potential shown in the literature have proven that ILs are very capable alternatives in the oil industry to fulfil the needs of chemical EOR.

1.3.3 Ionic Liquids in Unconventional Heavy Oil Recovery

Consisting of mainly extra heavy oil and sand bitumen, these oils typically require minimal processing after its extraction; however, the technology required in recovering them is highly specialized and can be expensive [28]. Unconventional oil recovery methods share similar problems often faced in EOR, in that the oils are often very viscous (particularly at lower temperatures) and high density [29]. The oil recoveries from heavy oil reservoirs involve thermal stimulation by direct heating or fluid injection, steam-based processes, and in situ combustion processes. These are some commonly used thermal-based methods [28].

Table 1.2 Ionic liquids for EOR from different applications [13]

Ionic liquids (Ils)	Major influencing factors	Type of reservoir rock	Porosity (%) and permeability (mD)	Additional oil recovery (%OOIP)
1-Hexadecyl-3-methyl imidazolium bromide [C$_{16}$MIM] $^+$[Br]$^-$	IFT reduction	Polysynthetic microchannel and sand pack	40–41% porosity for sandpack	0.2
1-Butyl-3-methylimidazolium chloride [BMIM]$^+$[Cl]$^-$, 1-butyl-3 methylimidazolium hydrogen sulphate [BMIM]$^+$[HSO$_4$]$^-$ 1-hexyl-3-methylimidazolium bromide [HMIM]$^+$[Br]$^-$	IFT reduction	Quartz	Not studied	Not studied
1-Methyl-3-alkylimidazolium bromide and 1-methyl-3-alkylimidazolium tetrafluoroborate	IFT reduction, phase behaviour.	Sand pack	38–39%, 1–2 Darcy	28–32% of OOIP
Poly ionic liquids based on 2-acrylamido-2-methylpropane	Wettability change and IFT reduction	Berea sandstone	20–22%, 180–300 mD	Additional 28% of OOIP

As it is natural of nonrenewable energies, sources for conventional oil will gradually decline. Therefore, focus has been shifted to unconventional heavy oil, shale oil and oil sand reserves to continually fulfil the demand that only oil-based fuels can deliver. While the total amount of unconventional resources in existence certainly exceeds the amount of conventional oil reserves, its gathering and extraction process is much more difficult and expensive. Furthermore, all known thermal extraction methods for this oil source carry the drawback of being environmentally harmful. As phase behaviour studies of ILs used in conventional EOR have shown success so in the field of heavy oil recovery ILs can be employed [30].

Chasib [31] experimented on using four types of reversible ionic liquids (RevILs) for the extraction of kerogen from Iraqi oil shale. RevILs are unique due to their ability to switch from into a liquid with molecular properties, allowing for the coupling reaction and separation process to take place within a single unit. This is advantageous as the kerogen is miscible with the molecular liquid, e.g. 3-aminopropyl)-triethoxysilane (TESA), while the RevIL counterpart 3-(triethoxysilyl)-propylammonium 3-(triethoxysilyl)-propyl carbamate (TESAC) was immiscible in kerogen. The author has also experimented with the use of multiple RevILs in a blend to obtain a unique binary mixture. Among the 4 ionic liquids tested, TESAC is reported to have the lowest viscosity, at under 1000 centipoise [32]. From a kerogen extraction test, TESAC was found to show the highest extraction efficiency of 90.2%, showing agreement with the referred literature, with 3-(trimethoxysilyl)-propylammonium 3-(trimethoxysilyl)-propyl carbamate (TMSAC) in second place, with roughly 80% efficiency. In kerogen extraction test of the binary RevIL mixtures, the best performing mixture of TESAC-TMSAC showed a tremendous recovery efficiency of 97.9%. While the results are very promising, the author has noted that its economic value is likely to be better only compared to the steam explosion technology, which is a thermal-based recovery method. Regardless of its current financial viability, the prospect of having an effective non-toxic alternative for use in heavy oil recovery certainly ensures RevIL to be an attractive choice.

Cao et al. [33] studied the different behaviours of various solvents, including nanofluids and ILs, for different heavy oil-coated surfaces to represent sandstone and carbonate solid surfaces. From an initial wettability analysis, ionic liquid 1-Butyl-2,3-dimethyl-imidazolium tetrafluoroborate (referred to as IL4) and the high-pH solution sodium metaborate ($NaBO_2$) generally gave the lowest contact angle results across various combinations of surfaces. Thus, the two solvents were considered for further testing under standard conditions. Both $NaBO_2$ and IL4 showed wettability altering capabilities by shifting the contact angle by 100° to roughly 50°. This change towards water-wet preference of the surface also shows an increase in oil recovery potential [34]. While $NaBO_2$ showed 34% OOIP recovery potential, IL4 showed 48% OOIP recovery, indicating of its ability to reverse the preferential affinity of the solid matrix from oil to water. It was also demonstrated that $NaBO_2$ and IL4 showed further reduction in contact angle with steadily increasing presence of heptane. When tested under differing temperatures of up to, the nanofluids were found to be thermally unsuited as precipitation occurred between 90 and 120 °C for two of the nanofluids tested; however, IL4 and $NaBO_2$ showed stability under high temperatures and pressures of up to 200 °C and 500 psi.

Cui et al. [35] investigated the effects of nanofluids SiO_2 and Al_2O_3, and the IL of $BMIMBF_4$ in altering the behaviour of a heavy mineral oil sample with heptane as the solvent. The results indicated that while the nanofluids are able to change the interfacial properties of the oil and aqueous layers, the IL was able to play the most efficient role for the purpose of residual oil saturation as part of the unconventional EOR process.

The use of DES ionic liquid mixtures in brine was further explored by Mohsen-zadeh et al. [36] for use in heavy oil recovery, using Berea sandstone as the surface medium in the contact angle and core flood tests. Both DESs have demonstrated good potential for heavy EOR, particularly at higher temperatures (as shown by the reduced IFT). The more alkaline of the two DESs have shown better performance in enhancing oil recovery. Of the water flooded residual heavy oil, between 14 and 30% oil was successfully recovered after the application of the DES solvents, averaging an OOIP% factor of around 70%.

The use of room temperature ILs was studied by Fan et al. [37] for its effects in reducing the viscosity of heavy oils. The ILs used is $[(Et)_3NH][AlCl_4]$, alongside several variations in which promoters of Ni^{2+}, Fe^{2+}, and Cu^+ was added. While the IL alone has shown to reduce the viscosity of the untreated oil from 420 mPa.s to 235 mPa.s alone, the use of ionic promoters manages to reduce it further down to as low as 148 mPa.s. From a compositional analysis, the asphaltene content decreased similarly, as supported by the reduction in molecular weight and sulphur content, due to the weakened C-S bonds. Thus, the use of ionic promoters too can show tremendous value in enhancing EOR performance for heavy oils.

Other tests of ionic liquid modifications were done by Shaban et al. [38], to examine their ability in reducing the viscosity of heavy oils, with further analysis of how the heavy crude oils has changed in sulfur content after the IL treatment. The IL tested is imidazolium chloride, [BMIM][Cl] modified with transition metals into a dual-functioning IL of imidazolium tetracholoferrate, [BMIM][FeCl4]. The modified IL was found to be thermally stable, displaying decompositions of roughly 20 wt% at 270 °C and 59 wt% at 380 °C. In terms of property alteration performance, the modified IL was able to obtain a slightly better viscosity reduction than the unmodified [BMIM][Cl]. Under different temperatures, the viscosity reduction induced by both ILs increases steadily with higher temperatures, though the optimal value is still found to be in between 70 °C and 90 °C. It was noted however that the presence of water can negatively impact the overall EOR performance of the IL. At a 0.1 wt% presence of water, the viscosity reduction and sulphur reduction achieved with introduction of IL into the crude oil is at 59% and 6.5%, respectively. However, at a mere 1 wt% presence of water, viscosity reduction and sulphur reduction shot down to 51% and 5.9%, respectively.

From the cases looked into, ILs have shown potential in extracting heavy oils as part of EOR, largely due to the similar mechanics in place required (wettability alteration, IFT and viscosity reduction, high thermal operation, etc.). However, the techniques and ILs explored by the literature are mostly different in that the ILs are modified in some manner to either accommodate for unique shale oil contents such as kerogen, varying surface types, and most importantly the presence of substances that can threaten the performance of the IL, such as presence of water or saline conditions. However, the results obtained through exploration of various types of ILs such as DESs, dual-functioning ILs, and RevILs demonstrate that the use of ILs as a suitable active solvent replacement can be a viable economical alternative, but more importantly one that is environmentally safe. Similar to the previous section, the noticeable gap within the research for heavy oil EOR ILs is its performance within a real field test environment.

1.4   Application of Ionic Liquids in Flow Assurance

While flow assurance is a relatively new field, its problems are very closely related to any type of issues that may disrupt nominal pipeline movement. In the oil and gas industry, the problems typically comprise presence of gas hydrates, wax and asphaltene depositions, scaling formation, corrosion, and slugging [34, 39–41]. For a large portion of these problems, they can be tackled by either use of thermal management methods or chemical additives, though they are certainly not limited to solely these two [42–44].

One such example among the chemical additives is of ionic liquids. While ionic liquids have not been given considerable attention as alternative to conventional solvents, it has shown tremendous potential in solving several specific issues in flow assurance. While it is largely difficult to simulate for, and thus requires careful laboratory experimentation, the dissolutive properties carried by ionic liquids have been applied for solving the problems listed above, particularly for the prevention of gas hydrates, wax, and asphaltene depositions. In fact, its viability is understood much clearer due to the more common issues associated with conventional solvents used, such as high toxicity to the environment, and thermal and pressure instabilities. Imidazolium-based and ammonium hydroxide-based ionic liquids show behaviours that make them the prime candidates for such applications. Its performance cannot be taken in isolation however due to the wide range of unexpected factors that may inhibit its performance, and thus more investigation will be required for most ionic liquids. The detailed discussion of application of ionic liquids in hydrate, corrosion mitigation, wax, scale and asphaltenes deposition control is done in upcoming chapters.

1.5 Conclusion and Future Application Prospects

As demonstrated from the cases, ionic liquids have shown to perform remarkably as drilling fluid additives and solvents for use in EOR. Its general propensity for functioning optimally within the higher temperatures and pressure conditions to be expected within subsurface conditions is their most salient point to consider over conventional solvents. The adaptability of ionic liquids within harsher conditions, such as within mixtures of other organic solvents or higher saline presence, shows that there is potentially an ionic liquid mixture that can be prepared for nearly all types of situations. It should be noted that certain ionic liquids can still be cost-prohibitive, which prevents it from being used in larger-scale applications. However, the closeness of performance within experiments that tests with multiple ILs have shown that there is still room to navigate in terms of selecting the most efficient material for further studies both in terms of performance and economic viability. Thus, the application of ILs can be expanded to the fields of heavy oil recovery, preparation of drilling fluid and as a chemical inhibitor in flow assurance employing chemical method.

The issue of conventional EOR surfactants not being stable at reservoir conditions, coupled with their hazardous environmental side effects, is the main motivations in necessitating the use of ILs as alternatives. In the light of this interest, development of simulation-based research tools to alleviate the inhibitive costs of IL research was done. Though relatively recent, one such example is the Conductor-Like Screening Model for Real Solvents (COSMO-RS) [45, 46], which have been used to identify the ionic properties for use in gas hydrate inhibitors [3, 47] and potential use in denitrification of liquid fuels [48].

The advantages of ILs can be plentiful in terms of its potential economic benefits. This can come in many forms, such as the lower dosage requirement compared to surfactants, which necessitates only a small amount per injection or application.

In the same vein, the larger temperature stability range could save costs through reduction in cycling of materials. The general superiority in oil yield performance will also contribute directly to economic gains for cases where high throughput or demand necessitates it. Even indirectly, in cases such as the need to minimize risks of environmental harm, financial decisions could be made without the need to allocate heavily for potential cleanup of hazardous wastes. Ultimately, the most important target to strive for in economic feasibility is to fully realize a scaled up process of the laboratory experiments. Therefore, there is certainly a gap that can be closed within the research, such as for larger field testing conditions, performance cost-benefit analysis compared to other surfactants, and further exploration of ionic liquid applications in the oil and gas industry.

References

1. Welton T (1999) Room-temperature ionic liquids. solvents for synthesis and catalysis. Chem Rev 99(1):2071–2083
2. Cláudio AFM, Swift L, Hallett JP, Welton T, Coutinho JAP, Freire MG (2014) Extended scale for the hydrogen-bond basicity of ionic liquids. Phys Chem Chem Phys 16(14):6593–6601
3. Khan MS, Bavoh CB, Partoon B, Lal B, Bustam MA, Shariff AM (2017) Thermodynamic effect of ammonium based ionic liquids on CO_2 hydrates phase boundary. J Mol Liq 238:533–539
4. Xiao Y, Malhotra SV (2004) Diels-Alder reactions in pyridinium based ionic liquids. Lett 45(45):8339–8342
5. Qasim A, Khan MS, Lal B, Shariff AM (2019) A perspective on dual purpose gas hydrate and corrosion inhibitors for flow assurance. J Pet Sci Eng 183:106418
6. Ingram T, Gerlach T, Mehling T, Smirnova I (2012) Extension of COSMO-RS for monoatomic electrolytes: Modeling of liquid-liquid equilibria in presence of salts. Fluid Phase Equilib 314:29–37
7. Kurnia KA, Lima F, Claudio AF, Coutinho JAP, Freire MG (2015) Hydrogen-bond acidity of ionic liquids: an extended scale. Phys Chem Chem Phys 17(29):18980–18990
8. Plechkova NV, Seddon KR (2008) Applications of ionic liquids in the chemical industry. Chem Soc Rev 37(1):123–150
9. Allen D et al (2002) An investigation of the radiochemical stability of ionic liquids. Green Chem 4(2):152–158
10. Bassane JFP et al (2016) Study of the effect of temperature and gas condensate addition on the viscosity of heavy oils. J Pet Sci Eng 142:163–169
11. Gateau P, Hénaut I, Barré L, Argillier JF (2004) Heavy oil dilution. Oil Gas Sci Technol 59(5):503–509
12. Zhu T, Walker JA, Liang J, Laboratory PD (2008) Evaluation of wax deposition and its control during production of Alaska North Slope oils, no. December
13. Bera A, Agarwal J, Shah M, Shah S, Vij RK (2020) Recent advances in ionic liquids as alternative to surfactants/chemicals for application in upstream oil industry. J Ind Eng Chem 82:17–30
14. Wilkes JS, Zaworotko MJ (1992) Air and water stable 1-ethyl-3-methylimidazolium based ionic liquids. J Chem Soc Chem Commun (13):965–967
15. Jacob N (2011) Israelachvili, intermolecular and surface forces, 3rd edn. Elsevier, Waltham
16. Fink J (2003) Oil Field Chemicals. Elsevier
17. Yang L, Jiang G, Shi Y, Yang X (2017) Application of ionic liquid and polymeric ionic liquid as shale hydration inhibitors. Energy Fuels 31(4):4308–4317

18. Luo Z, Wang L, Yu P, Chen Z (2017) Experimental study on the application of an ionic liquid as a shale inhibitor and inhibitive mechanism. Appl Clay Sci 150:267–274
19. Jia H et al (2019) Investigation of inhibition mechanism of three deep eutectic solvents as potential shale inhibitors in water-based drilling fluids. Fuel 244:403–411
20. Ofei TN, Bavoh CB, Rashidi AB (2017) Insight into ionic liquid as potential drilling mud additive for high temperature wells. J Mol Liq 242:931–939
21. Bera A, Belhaj H (2016) Ionic liquids as alternatives of surfactants in enhanced oil recovery-A state-of-the-art review. J Mol Liq 224:177–188
22. Nandwani SK, Malek NI, Lad VN, Chakraborty M, Gupta S (2017) Study on interfacial properties of imidazolium ionic liquids as surfactant and their application in enhanced oil recovery. Colloids Surf A Physicochem Eng Asp 516:383–393
23. Dong B, Zhao X, Zheng L, Zhang J, Li N, Inoue T (2008) Aggregation behaviour of long-chain imidazolium ionic liquids in aqueous solution: micellization and characterization of micelle microenvironment. Colloids Surf A Physicochem Eng Asp 317(1–3):666–672
24. Vanyúr R, Biczók L, Miskolczy Z (2007) Micelle formation of 1-alkyl-3-methylimidazolium bromide ionic liquids in aqueous solution. Colloids Surf A Physicochem Eng Asp 299(1–3):256–261
25. Velusamy S, Sakthivel S, Sangwai JS (2017) Effect of imidazolium-based ionic liquids on the interfacial tension of the alkane-water system and its influence on the wettability alteration of quartz under saline conditions through contact angle measurements. Ind Eng Chem Res 56(46):13521–13534
26. Nabipour M, Ayatollahi S, Keshavarz P (2017) Application of different novel and newly designed commercial ionic liquids and surfactants for more oil recovery from an Iranian oil field. J Mol Liq 230:579–588
27. Abdullah MMS, AlQuraishi AA, Allohedan HA, AlMansour AO, Atta AM (2017) Synthesis of novel water soluble poly (ionic liquids) based on quaternary ammonium acrylamidomethyl propane sulfonate for enhanced oil recovery. J Mol Liq 233:508–516
28. Speight JG (2013) Enhanced recovery methods for heavy oil and tar sands
29. Speight JG (1998) Asphaltenes and the structure of petroleum. Oil Gas Sci Technol 59(5):467–477
30. Sakthivel S, Velusamy S, Gardas RL, Sangwai JS (2015) Nature friendly application of ionic liquids for dissolution enhancement of heavy crude oil. Proc - SPE Annu Tech Conf Exhib 2015–January, pp 6583–6594
31. Chasib KF (2017) Extraction of kerogen from oil shale using mixed reversible ionic liquids, no. JIChEC
32. Blasucci VM, Hart R, Pollet P, Liotta CL, Eckert CA (2010) Reversible ionic liquids designed for facile separations. Fluid Phase Equilib 294(1–2):1–6
33. Cao N, Mohammed MA, Babadagli T (2017) Wettability alteration of heavy-oil-bitumen-containing carbonates by use of solvents, high-pH solutions, and nano/ionic liquids. SPE Reserv Eval Eng 20(2):363–371
34. Khan MS, Lal B, Keong LK, Ahmed I (2019) Tetramethyl ammonium chloride as dual functional inhibitor for methane and carbon dioxide hydrates. Fuel 236:251–263
35. Cui J, Babadagli T (2017) Use of new generation chemicals and nano materials in heavy-oil recovery: Visual analysis through micro fluidics experiments. Colloids Surf A Physicochem Eng Asp 529:346–355
36. Mohsenzadeh A, Al-Wahaibi Y, Jibril A, Al-Hajri R, Shuwa S (2015) The novel use of deep eutectic solvents for enhancing heavy oil recovery. J Pet Sci Eng 130:6–15
37. Fu Fan H, Bao Li Z, Liang T (2007) Experimental study on using ionic liquids to upgrade heavy oil. Ranliao Huaxue Xuebao/J Fuel Chem Technol 35(1):32–35
38. Shaban S, Dessouky S, Badawi AEF, El Sabagh A, Zahran A, Mousa M (2014) Upgrading and viscosity reduction of heavy oil by catalytic ionic liquid. Energy Fuels 28(10):6545–6553
39. Khan MS, Cornelius BB, Lal B, Bustam MA (2018) Kinetic assessment of tetramethyl ammonium hydroxide (ionic liquid) for carbon dioxide, methane and binary mix gas hydrates. In: Rahman MM (ed) Recent advances in ionic liquids. IntechOpen, London, UK, pp 159–179

40. Cornelius BB, Lal B, Khan MS, Osei H, Ayuob M (2018) Inhibition effect of 1-ethyl-3-methylimidazolium chloride on methane hydrate equilibrium. J Phys Conf Ser 1123:012060
41. Lal B, Nashed O. (2019) Chemical additives for gas hydrates, 1st ed. Springer International Publishing
42. Qasim A, Khan MS, Lal B, Ismail MC, Rostani K (2020) Quaternary ammonium salts as thermodynamic hydrate inhibitors in the presence and absence of monoethylene glycol for methane hydrates. Fuel 259, no. July 2019, p 116219
43. Sivabalan V, Walid B, Madec Y, Qasim A, Lal B (2020) Corrosion inhibition study on glycerol as simultaneous gas hydrate and corrosion inhibitor in gas pipelines. Malaysian J Anal Sci 24(1):62–69
44. Khan MS et al (2018) Experimental equipment validation for methane (CH_4) and carbon dioxide (CO_2) hydrates. IOP Conf Ser Mater Sci Eng 344:1–10
45. Klamt A, Eckert F, Arlt W (2010) "COSMO-RS - An alternative to simulation for calculating thermodynamic properties of liquid mixtures. Annu Rev Chem Biomol Eng 1(1):101–122
46. Bavoh CB et al (2016) COSMO-RS: an ionic liquid prescreening tool for gas hydrate mitigation. Chinese J Chem Eng 24(11):1619–1624
47. Sulaimon AA, Tajuddin MZM (2017) Application of COSMO-RS for pre-screening ionic liquids as thermodynamic gas hydrate inhibitor. Fluid Phase Equilib 450:194–199
48. Hizaddin HF, Hashim MA, Anantharaj R (2013) Evaluation of molecular interaction in binary mixture of ionic liquids + heterocyclic nitrogen compounds: Ab initio method and cosmo-rs model. Ind Eng Chem Res 52(50):18043–18058

Chapter 2
Application of Ionic Liquids in Gas Hydrate Inhibition (GHI)

2.1 Introduction

Gas hydrates are crystallized solid compounds in which gas molecules form hydrogen bonds with water and are trapped inside. Gas hydrates are formed at low temperature and high pressure conditions. The bond is sustained through Wan der Waals forces [1, 2]. Some hydrate forming gases include carbon dioxide, methane, ethane, propane and hydrogen sulphide under production and conveyance situations [3].

Flow assurance is a critical task in the oil and gas industry, which requires proper management to assure successful operation of the flowline. In deep-sea systems, flow control issues are becoming more prevalent and need to be effectively addressed [4]. Once such issue is the formation of gas hydrates, which poses a serious threat to oil and gas transmission pipeline integrity and carries significant economic loss and safety risks [1, 5]. As the hydrate formation process materializes in the flowline, the interaction occurs between the hydrate formed and the pipeline at each stage of the process. The pipeline industry suffers losses of roughly US$ 1 million per day due to hydrate plugging [6]. To tackle this problem, a variety of hydrate inhibition techniques were used, which proved to be uneconomical in most cases. Many of these practices face the challenges involving regeneration, requirement of large storage space, and effective biodegradability, among others [7, 8].

Several approaches that were implemented to avoid the aforementioned pipeline issues include dehydration, heating, chemical and mechanical methods [9]. Apart from the chemical process, the methods listed are either ineffective or involve a significant amount of chemical solvents. These methods raise the operational and capital expenditure in maintaining the oil and gas facilities and have a negative effect on the climate [10, 11]. This renders the chemical process as the only feasible alternative for the pipeline industry. In order to mitigate hydrate formation in the pipeline, the flow assurance industry has used many different types of organic and inorganic solvents, though the use of these compounds requires huge infrastructure and workspace which makes their utilization economically unfeasible [12]. To address this issue, research

© The Author(s), under exclusive licence to Springer Nature Switzerland AG 2021   17
B. Lal et al., *Ionic Liquids in Flow Assurance*,
SpringerBriefs in Petroleum Geoscience & Engineering,
https://doi.org/10.1007/978-3-030-63753-8_2

focus has been placed on environmentally benign compounds for handling the issue of flow assurance. These compounds can prove to be cost-effective.

2.2   Important Concepts of Hydrate Formation

2.2.1   Formation of Gas Hydrates

Within a flow pipeline where oil and gas are transported, there exists typically three phases within them. These three phases are identified as gas, hydrocarbon and liquid. The flow model of hydrate development is classified into four types. The first type is categorized as the oil dominated system. The second is the gas-dominated system, and it contains lesser amount of hydrocarbon liquid. This kind of system does not usually cause hydrate blockage. The third type is the condensate system. Its peculiarity is in the droplet formation in the system, caused by the suspension of the dissolved water within the condensate. High-water-cut system is the fourth type among these systems. In this system, the volume may reach up to 70% water cut, and water becomes immiscible in the oil. The temperature varies from 275 to 285 K while the pressure lies in the range of 3–10 MPa in this system [3] and is non-stochiometric. These compounds have a polyhedral structure through hydrogen bonding. It is kept stable through Wan der Waals forces. Figure 2.1 demonstrates the obstruction caused by gas hydrates inside the pipeline [13].

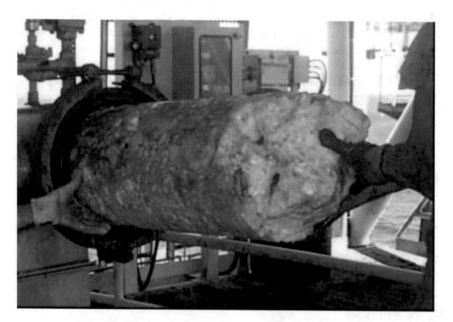

Fig. 2.1 Hydrate plugging inside pipeline [13]

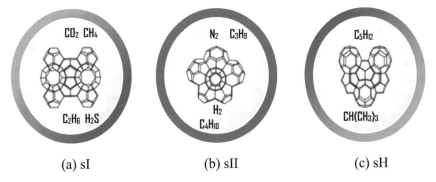

Fig. 2.2 Structure types of gas hydrates sI (**a**), sII (**b**), and sH (**c**)

2.2.2 Gas Hydrate Structures

Gas hydrates are categorized as structure I (sI), structure II (sII) and structure H (sH) in nature. The structure consists in a repeated form. The hydrate structures are shown in Fig. 2.2a–c). Structure I and structure II hydrates are cubic in shape while structure H hydrates are hexagonal. For structure I, the diameter range of guest molecules is between 4.2 and 6 Å. Methane (CH_4), ethane (C_2H_6), hydrogen sulphide (H_2S) and carbon dioxide (CO_2) gases form structure I hydrates. Guest molecules in Structure II hydrates are of sizes less than 4.2 Å diameter. The gases which form structure II hydrates as guest molecules are nitrogen (N_2), propane (C_3H_8), hydrogen (H_2) and butane (C_4H_{10}). Molecules of relatively bigger chain length form structure H hydrates which usually lie in the range of 7–9 Å. The compounds which form structure H hydrates do so by combining with smaller molecules. Gases like pentane and isobutane form structure H in combination with methane, nitrogen and hydrogen sulphide which are relatively smaller in size [14]. The structure H hydrates are considerably different as compared to structure I and structure II hydrates but they are seldom found in nature [6].

2.3 Gas Hydrate Inhibition

The development of gas hydrates inside flow pipelines can be prevented by many traditional techniques such as heating, dehydration, chemical addition and depressurization. All these methods have their own advantages and disadvantages depending on the situation encountered [15]. In the depressurization method, pressure is reduced from one side of the flowline when the hydrate forms a plug. However, this technique is not suitable for oil and gas pipelines with high pressure and large distances due to safety concerns [6]. Additionally, it may cause hydrate plug velocity to increase inside the pipeline which damages its bending and vents thus compromising the overall flow integrity [16]. For the heating method, steam is used to heat the pipeline in order to keep the temperature conditions within the pipeline outside of hydrate forming regions, thus keeping the operation normal [17]. However, this process has proved to be highly uneconomical as it requires about US$ 1 million per km of flowline length to work effectively [3]. Dehydration or water removal has been recommended as a successful preventive method in the literature [18]. Hydrate formation will cease to occur if the water content is unavailable, but due to the labour costs in human involvement, it becomes economically infeasible. This scheme is still useful in downstream applications and other gas processing plants [19].

Hydrate management can also be done by adding compounds that push the phase boundaries and keep the pressure and temperatures out of the hydrate occurring zone and avoid its formation. These compounds avert hydrate formation by either delaying it or changing the phase behaviour of the process [20, 21]. Among these strategies, inhibition of hydrate formation by use of chemicals is currently favoured by the oil and gas industry. Chemical inhibitors do not show adverse effect towards normal operation of the pipeline and are considered safe to use [22]. In terms of classification, thermodynamic hydrate inhibitors come under the category of high dosage hydrate inhibitors while kinetic hydrate inhibitors (KHIs) and anti-agglomerants (AAs) are classified as low dosage hydrate inhibitors (LDHIs). Their classification is shown in Fig. 2.3. THIs prevent hydrate formation by shifting the hydrate-liquid-vapour equilibrium (HLVE) curve towards low temperature and high-pressure domains, thus avoiding conditions of hydrate formation [23, 24]. KHIs avert the formation of hydrates by means of slowing nucleation for a certain amount of time, i.e. it is prolonged as long as free water phase is present inside the oil and gas pipeline, thereby avoiding it [25, 26]. Whereas AAs disallow formation of a solid plug inside the pipeline by only allowing smaller lumps of hydrates to form. These small chunks of hydrate do not cause performance hindrances in pipeline flow [27, 28]. Considering the above-mentioned discussion, it is argued that the use of chemical inhibitors is a preferable defensive strategy to be used inside flowlines. While it is considered the most economical method, further merits of this strategy will be discussed in upcoming sections.

In flow assurance applications, several compounds have already established itself for conventional use. In KHI application, Polyvinylcaprolactum (PVCap) and Polyvinylpyrrolidone (PVP) are the foremost chemicals to be employed in the oil and gas industry [29–31]. Regarding thermodynamic inhibition, methanol (MeOH) and monoethylene glycol (MEG) are the most commonly used THIs which have worked successfully under pipeline conditions [32–34]. However, the usage of these compounds is limited to either KHI or THI applications only, which restricts their utilization. Researchers are focusing on inhibitors which can address the issue of gas hydrate and corrosion inhibition thus acting as dual-purpose mitigators [35, 36]. Ionic liquids in combination with some amino acids are the commonly used chemicals employed in dual functional hydrate inhibition applications [32, 37, 38].

Ionic liquids having imidazolium [15, 39, 40] and ammonium bases [35, 41, 42] are employed in THI and KHI applications in the literature. These compounds have thermal stability and low electric conductivity [43]. Imidazolium-based ILs consist of cationic groups of ethyl-methyl-im (C_2-C_1-im) and propyl-methyl-im (C_3-C_1-im).

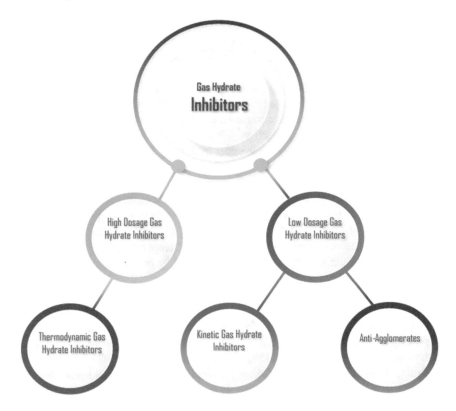

Fig. 2.3 Gas hydrate inhibitors classification (GHIs)

The anionic groups of imidazolium-based ILs included bromine (Br), chlorine (Cl) and iodine (I) among many others [44, 45]. For ammonium-based ILs, tetramethyl (TMA), tetraethyl (TEA) and tetrapropyl (TPA) are some of the cations employed in research. These cations have mostly been engaged with the hydroxide, bromide, chloride anions and others [46, 47]. Various research efforts in the past have employed a host of anions and cations to investigate thermodynamic inhibition capabilities of these chemicals [48, 49]. In a previous study, It was recommended by Tariq et al. [49] to use OH, NH$_2$ and SO$_3$H functional groups to investigate their hydrogen bonding capabilities.

2.3.1 Thermodynamic Hydrate Inhibitors

As discussed before, THIs are used in bulk quantities in the flowline to prevent gas hydrate formation. These compounds avoid hydrate formation regions by shifting the hydrate liquid vapour equilibrium towards a condition of high pressure and low temperature [50, 51]. The hydrate mitigation capability of methanol and mono ethylene glycol depends on the attraction between O_2 atoms of the inhibitor and the water molecules [35]. Due to the polarity of inhibitor molecules, they form hydrogen bond with molecules of water. The presence of hydrogen atom can explain the mechanism wherein electrons are migrated to the other atom which is then bonded covalently. This exposes the positively charged hydrogen proton. This proton in turn bonds electrostatically with other electrons and forms an attraction between negative charged molecules. In order to achieve effective inhibition performance, the water-inhibiting hydrogen bond should be stronger than water itself [52]. Safety cost of operation and physical properties are some of the factors taken into consideration while selecting THIs [34]. Along with these features, regeneration and reinjection capabilities of the used chemical are also an important step in the selection process [53, 54]. THIs are typically applied in higher concentrations for gas hydrate inhibition, in the range of 5–50 by weight percentage (wt%). The inhibitor concentration used is specified on the basis of operational conditions and surrounding temperatures.

Methanol and monoethylene glycol are some of the organic compounds which have commercially been applied as THIs [55–57]. Cha et al. [56] investigated the THI performance of monoethylene glycol at 10, 30, and 50 wt% for methane gas hydrates. Ng et al. [58] studied the THI inhibition of methanol in the case of methane hydrates. Moreover, Anderson et al. [59] also examined the performance of methanol in shifting the equilibrium for methane hydrates and their results are in agreement with previous studies by Ng et al. [58].

Commercial THIs are currently being used for hydrate inhibition purposes but due to their high volatility their use can be unsafe in some situations. Along with volatility issues, these compounds have not shown environmentally friendly behaviour [8]. In consideration of the environment, the pursuit of new 'green' THIs is essential, thus prompting research in new biodegradable compounds in ionic liquids that can play a significant role in lessening environmental harm.

For THI application, many different ionic liquids have been used. Mostly the studies have been performed employing ionic liquids having imidazolium present in the cationic moeity. The HLVE behaviour of these ionic liquids has been studied in the presence of carbon dioxide and methane [49, 60]. Ionic liquids belonging to other families can also be studied. In this regard, Yaqub et al. recommended further investigation of ammonium and piperidinium base ionic liquids to probe for their phase behaviour, especially in the case of natural gases [8]. Butyl methyl imidazolium has been used in previous researches among imidazolium-based ionic liquid compounds [36, 61] in combination with tetrafluoroborate [15, 62], chloride [61, 63], bromide, iodide, methyl sulphate, trifluoromethanesulfonate, and dimethyl sulphate anions. These compounds have been used for methane and carbon dioxide hydrate inhibition in KHI and THI applications. Likewise, ethyl methyl imidazolium tetrafluoroborate, ethyl methyl imidazolium chloride, ethyl methyl imidazolium bromide and ethyl methyl imidazolium trifluoromethanesulfonate have also been used as methane and carbon dioxide hydrate inhibitors in THI and KHI applications. For piperidinium-based ionic liquids, ethyl methyl piperidinium bromide and ethyl methyl pipredenium tetrafluoroborate have been used as THIs to suppress carbon dioxide hydrates

[36]. For CO_2 hydrate control, ethyl methyl morpholinium bromide and ethyl methyl morpholinium tetrafluoroborate ionic liquids have also been employed [44, 51].

2.3.2 Low Dosage Hydrate Inhibitors

The performance of LDHIs as hydrate inhibitors has been studied considerably in the literature [54]. Kinetic hydrate inhibitors (KHIs) and anti-agglomerants (AAs) are types of low dosage hydrate inhibitors.

2.3.2.1 Kinetic Hydrate Inhibitors (KHIs)

KHIs prevent hydrate development through postponement. These inhibitors evade nucleation of the free water phase in the pipeline in excess of the time of residence [54, 64]. The nucleation time of hydrate formation is referred to as induction time. It marks the time interval between equilibrium state to complete hydrate crystal appearance [65]. At this condition, their properties are stabilized. KHIs are usually employed in concentration range of 0.5–2 wt% which are considered to be lower [66]. The performance of KHI is considered time-dependent, contrasting to that of THIs [29]. Deeper systems are more prone to falling into the hydrate region as compared to the shallow systems. The deeper systems require KHIs to deal with the threat of hydrate formation [54, 67]. KHIs prevent growth and nucleation of hydrate crystals by adsorbing to the surface [12]. Due to its applicability for a wide range of hydrocarbon systems, KHIs have grown to widespread commercial use within the last two decades [68].

Polyvinylcaprolactam (PVCap) and polyvinylpyrrolidone (PVP) are examples of commercially used kinetic hydrate inhibitors. These compounds are readily water soluble [20, 65, 69]. PVP and PVCap consist of polyethylene constituents attached with N atom and $C = O$ group chemical rings which are polar and sphere-shaped [68, 70]. In hydrate crystal cage, these KHI polymers attach through adsorption at the surface [65].

Studies in which ionic liquids are employed as KHI have shown recent growth [44, 49]. Nashed et al. [71] studied the KHI performance of imidazolium-based ILs including (C_4C_1im) (CF_3SO3), (C_4C_1im) (CH_3SO_4) and $(OH-C_2C_1im)$ (Br). Results show that $(OH-C_2C_1im)$ (Br) expanded the induction time by almost double to that of pure water at 7.1 MPa and 258.15 K. In a recent study by Saad et al. [72], some new ammonium-based ionic liquids including tetramethyl ammonium hydroxide (TMAOH), tetraethyl ammonium hydroxide (TEAOH), tetrapropyl ammonium hydroxide (TPrAOH) and tetrabutyl ammonium hydroxide (TBAOH) were examined for thermodynamic and kinetic hydrate inhibition studies. The above-mentioned ionic liquids were studied for carbon dioxide and methane mixed gas hydrates.

Research continues to also grow for environmentally friendly hydrate inhibitors, and in this regard the application of natural biopolymers can prove to be a resourceful addition. Biopolymers are biodegradable compounds, so their use can reduce much of the harm done to the environment. Chitosan [8, 73] and anti-freezing proteins [54, 74] are some of the biopolymers that have been considered as feasible hydrate inhibitors. KHI study of starch has been performed among biomolecules. Amylopectin is the basic ingredient of starch and starch is a mixture of linear polymer amylase and high branch amylopectin [75]. Yang et al. investigated the delay in induction time and amount of gas consumed for polyethylene and polypropylene in presence of starch. The effect was studied for methane, ethane, propane and butane in the pressure range of 1 to 8 MPa and temperature range of 275–280 K. Another biodegradable polymer examined for KHI behaviour is Chitosan [73, 76]. It is a polysaccharide composed of N-acetyl-D-glucosamine and type β-(1–4)-linked D-glucosamine [77]. Chitosan was found to be a better inhibitor and reduced hydrate formation time of methane and methane-ethane hydrates [8]. The study found that chitosan delayed the induction time for 2 h at 4.50 MPa and 274.3 K [73]. Pure water showed an induction time of only roughly 0.05 h under similar experimental conditions. Antifreeze proteins (AFPs) belong to a new class of natural KHIs [54, 78]. AFPs, which are obtained from plants and bacteria, are actually polypeptides and can survive in freezing conditions. Perfeldt et al. [79] studied the effect of AFPs obtained from Rhagium mordax, a class of beetles, on methane hydrate formation. It showed KHI capabilities by delaying the hydrate formation of methane. These compounds showed results similar to that of PVP.

Efforts focused towards finding a single inhibitor that can serve in the application of gas hydrate and corrosion inhibition; referred as a dual-purpose inhibitor or gas hydrate corrosion inhibitor (GHCI), have also grown as it can address the two major issues of flow assurance simultaneously [80]. For this purpose, some have used corrosion inhibitors belonging to polysynthetic groups (recognized for their strong interactions with the metal surface) in combination with basic polymers. The interaction between metal surfaces and the cations in the group is weaker, so it projects away from the surface when combined with the corrosion inhibitor groups. Park et al. [81] used the hydrate and corrosion inhibiting groups together as Vcap and PVCap were combined with quaternary ammonium corrosion inhibiting groups of taurine and imidazole. The use of copolymers for gas hydrate and corrosion inhibition application is an emerging area of interest in oil and gas industry.

2.3.2.2 Anti-Agglomerants (AAs)

Anti-agglomerant compounds are the type of LDHIs which break and disperse hydrates into small particles. These chemicals do not avert hydrate formation, as opposed to THIs and KHIs. Anti-agglomerants (AAs) can be classified as a surface active chemical, and in their presence fluid viscosity remains on the lower side, which allows hydrates to be transported in a state of slurry [82]. AAs adsorb on hydrated droplet surfaces and transform water into hydrates. The head group of the anti-agglomerant compound is attached and in fluid phase, the hydrate particle remain dispersed throughout the longer hydrocarbon part [83]. Their performance is time independent, and in some deep water conditions, the application of AAs is preferred over KHIs [74].

Longer-chain ILs that have surfactants are typically AAs, as they contain both hydrophilic and hydrophobic moieties. The attraction between hydrophilic end of ionic liquid which is anion and hydrogen ion of water molecule due is dipole-dipole interaction. The cationic part of ionic liquid which is hydrophobic dissolves in the gas oil or oil phase inside the flowline [68]. Quaternary ammonium compounds from n-butyl, iso-pentyl and n-pentyl groups were used as anti-agglomerants. One of such compound is tetrapentylammonium bromide (TPAB), which was found and tested by Shell Oil Company to have highly effective anti-agglomerating performance due to its hydrate dispersion properties [68]. Researchers from Dutch Shell previously used tetraalkylammonium salts (notable for their twin-tailed structure) as an anti-agglomerant; however, it is no longer used due to environmental concerns [84]. Span 20, Span 40, Span 60 and Span 80 are some of the commercially used surfactants as anti-agglomerants utilized by Huo et al. [85]. The Span compounds were able to keep the hydrate particles dispersed up to 8.2 MPa and 277 K. The performance of these compounds was also compared against the synthesized anti-agglomerant named dodecyl-2-(2-caprolactamyl) ethanamide. At high water-cut systems of up to 75%, the synthesized chemicals, however, displayed better dispersion. Kelland et al. directed their research towards the development of economical LDHIs using prolyproxylate compounds as anti-agglomerants for flow assurance applications [86]. It was reported that polyamine polypropoxylates and other branched polypropoxylates dispersed gas hydrates in a hydrocarbon fluid in well-agitated conditions.

2.4   ILs as Gas Hydrate Inhibitors

There are several important factors to consider in designing and selecting an effective hydrate inhibitor, of which are illustrated in Fig. 2.4. Selection of an appropriate alkyl chain length is the most important factor, as a shorter chain length is usually preferred in gas hydrate inhibition due to fact that that a shorter chain length is more suitable for stronger hydrogen bonding. Concentration of the inhibitor also plays a significant role in hydrate control, especially in THI applications. A higher concentration of inhibitor shows better result in terms of shifting the HLVE towards higher pressure and lower temperature regions. Furthermore, charge density and hydrogen bonding ability of the compound influences its ability to make hydrogen bonds which can produce a stronger inhibition effect [87, 88]. Finally, the ability of a gas hydrate inhibitor to adsorb on the surface is another key parameter, particularly for the selection of KHIs [40].

Fig. 2.4 Factors affecting
the gas hydrate inhibition

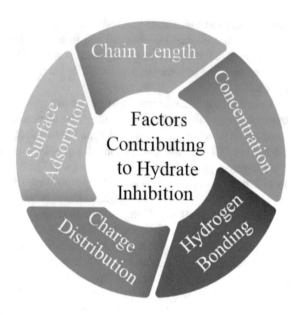

Through the cases studied, it was found that the factors above hold a large influence over the gas hydrate inhibition properties of ILs. Table 2.1 shows the performance of some ILs as gas hydrate inhibitors, studied for their potential application in oil and gas flow pipelines. The hydrate mitigation results of mostly imidazolium-based ILs alongside some alkyl hydroxide compounds have been examined in preceding literature, and THI and KHI results are mentioned in Table 2.1.

2.5 Summary

In this chapter, the role of ionic liquids as gas hydrate inhibitors is studied for their potential use in offshore pipelines. It was found that imidazolium, alkyl hydroxide, ammonium-based and quaternary ammonium salts (QAS) have been utilized for the application. THI and KHI studies of Ionic liquids for hydrate mitigation purposes are available in preceding literature. It was also found from the literature that the IL compound 1-ethyl-3-methylimidazolium ethyl sulphate $[EMIM]^+[EtSO_4]^-$ achieved better results in terms of hydrate inhibition, when compared to the other compounds within this scope of analysis. It was revealed that the main factor in gas hydrate inhibition for THIs is the ability of ILs to make hydrogen bonds. For most ILs, chain length, charge density of anionic and cationic moieties are the main factors affecting its performance. In closing, many ionic liquids have shown promising characteristics of hydrate mitigation, and their application in this field of research can be explored further.

Table 2.1 Gas hydrates inhibition (GHI) of Ionic Liquids

Ionic Liquids	Structure	GHI specifications (CH₄)	Results		References
			THI (K)	KHI (RIP)	
[EMIM]⁺[Cl]⁻		KHI+THI	1.22 (10 wt%)	3.08 (1 wt%)	[87]
[EMIM]⁺[BF₄]⁻		KHI+THI	0.52 (10wt%)	17 (0.1 wt%)	[15]
[BMIM]⁺[Br]⁻		KHI+THI	0.58 (10wt%)	5.8 (1 wt%)	[87]
[EMIM]⁺[EtSO₄]⁻		KHI+THI	2.15 (10wt%)	0.55 (1 wt%)	[15]
[BMIM]⁺[Cl]⁻		KHI+THI	0.69 (10wt%)	4.13 (1 wt%)	[87]
[BMIM]⁺[PF₆]⁻		THI	–	–	[89]
[TMA]⁺[OH]⁻		THI+KHI	1.53 (10wt%)	0.41 (1 wt%)	[11]
[TMA]⁺[Cl]⁻		THI+KHI	1.42 (10wt%)	0.063 (1 wt%)	[35]
[TEA]⁺[OH]⁻		THI+KHI	1.12 (10wt%)	0.85 (1 wt%)	[89]
[TPrA]⁺[OH]⁻		THI+KHI	0.66 (10wt%)	0.89 (1 wt%)	[89]

References

1. Zerpa LE, Salager J-L, Koh CA, Sloan ED, Sum AK (2011) Surface chemistry and gas hydrates in flow assurance. Ind Eng Chem Res 50(1):188–197
2. Yarveicy H, Ghiasi MM, Mohammadi AH (2018) Determination of the gas hydrate formation limits to isenthalpic Joule – Thomson expansions. Chem Eng Res Des 132:208–214
3. Sloan D et al (2010) Natural gas hydrates in flow assurance. Gulf Professional Publishing
4. Joshi SV et al (2013) Experimental flowloop investigations of gas hydrate formation in high water cut systems. Chem Eng Sci 97:198–209
5. Raja PB et al (2016) Reviews on corrosion inhibitors: a short view. Chem Eng Commun 203(9):1145–1156
6. Sloan ED et al (2008) Clathrate hydrates of natural gases, 3rd edn, vol. 87, no. 13–14. CRC Press Taylor & Francis, Boca Raton, London, New york
7. Khan MS, Lal B, Partoon B, Keong LK, Bustam MA, Mellon NB (2016) Experimental evaluation of a novel thermodynamic inhibitor for CH_4 and CO_2 hydrates. Procedia Eng 148(2016):932–940
8. Yaqub S, Lal B, Partoon B, Mellon NB (2018) Investigation of the task oriented dual function inhibitors in gas hydrate inhibition: a review. Fluid Phase Equilib 477:40–57
9. Yarveicy H, Ghiasi MM (2017) Modeling of gas hydrate phase equilibria: extremely randomized trees and LSSVM approaches. J Mol Liq 243:533–541
10. Sabil KM, Nashed O, Lal B, Ismail L, Japper-Jaafar A (2015) Experimental investigation on the dissociation conditions of methane hydrate in the presence of imidazolium-based ionic liquids. J Chem Thermodyn 84:7–13
11. Khan MS, Partoon B, Bavoh CB, Lal B, Mellon NB (2017) Influence of tetramethylammonium hydroxide on methane and carbon dioxide gas hydrate phase equilibrium conditions. Fluid Phase Equilib 440:1–8
12. Kelland MA (2014) Production Chemicals for the Oil and Gas Industry, 2nd edn. CRC Press, New York
13. Obanijesu EO (2012) Corrosion and hydrate formation in natural gas pipelines. Curtin University, Australia
14. Atilhan M, Aparicio S, Benyahia F, Deniz E (2012) Natural Gas Hydrates. In: Al-Megren H (ed) Advances in natural gas technology, vol. 1. InTech Janeza Trdine 9, 51000 Rijeka, Croatia, pp 194–212
15. Xiao C, Adidharma H (2009) Dual function inhibitors for methane hydrate. Chem Eng Sci 64(7):1522–1527
16. Drive C, Carroll J (2009) Natural gas hydrates: a guide for engineers, 3rd edn., no. October. Elsevier
17. Mohammad-Taheri M, Zarringhalam Moghaddam A, Nazari K, Gholipour Zanjani N (2013) The role of thermal path on the accuracy of gas hydrate phase equilibrium data using isochoric method. Fluid Phase Equilib 338:257–264
18. Gonfa G, Bustam MA, Sharif AM, Mohamad N, Ullah S (2015) Tuning ionic liquids for natural gas dehydration using COSMO-RS methodology. J Nat Gas Sci Eng 27:1141–1148
19. Perry RH, Green DW (2008) Perry's chemical engineers' handbook, 8th edn., vol. 1. McGraw-Hill Companies, Inc
20. Kamal MS, Hussein IA, Sultan AS, Von Solms N (2016) Application of various water soluble polymers in gas hydrate inhibition. Renew Sustain Energy Rev 60:206–225
21. Zare M, Haghtalab A, Ahmadi AN, Nazari K, Mehdizadeh A (2015) Effect of imidazolium based ionic liquids and ethylene glycol monoethyl ether solutions on the kinetic of methane hydrate formation. J Mol Liq 204(04):236–242
22. Wang F, Zhang JN, Li CX (2013) Study on hydrate inhibitor to prevent freeze-plugging of gas-condensate well. Adv Mater Res 868:737–741
23. Acosta HY, Bishnoi PR, Clarke Ma (2011) Experimental measurements of the thermodynamic equilibrium conditions of tetra-n-butylammonium bromide semiclathrates formed from synthetic landfill gases. J Chem Eng Data 56(1):69–73

24. Akhfash M, Arjmandi M, Aman ZM, Boxall JA, May EF (2017) Gas hydrate thermodynamic inhibition with MDEA for reduced MEG circulation. J Chem Eng Data 62(9):2578–2583
25. Khan MS, Cornelius BB, Lal B, Bustam MA (2018) Kinetic assessment of tetramethyl ammonium hydroxide (ionic liquid) for carbon dioxide, methane and binary mix gas hydrates. In: Rahman MM (ed) Recent advances in ionic liquids. IntechOpen, London, UK, pp 159–179
26. Makogon YF, Holditch SA, Makogon TY (2007) Natural gas-hydrates - A potential energy source for the 21st Century. J Pet Sci Eng 56(1–3):14–31
27. Braniff MJ (2013) Effect of dually combined under-inhibition and anti-agglomerant treatment on hydrate slurries
28. York JD, Firoozabadi A (2008) Comparing effectiveness of rhamnolipid biosurfactant with a quaternary ammonium salt surfactant for hydrate anti-agglomeration. J Phys Chem B 112(3):845–851
29. Nashed O, Partoon B, Lal B, Sabil KM, Mohd A (2018) Review the impact of nanoparticles on the thermodynamics and kinetics of gas hydrate formation. J Nat Gas Sci Eng 55:452–465
30. Mady MF, Kelland MA (2013) Fluorinated quaternary ammonium bromides: studies on their tetrahydrofuran hydrate crystal growth inhibition and as synergists with polyvinylcaprolactam kinetic gas hydrate inhibitor. Energy Fuels 27(9):5175–5181
31. Tonelli D, Capicciotti CJ, Doshi M, Ben RN (2015) Inhibiting gas hydrate formation using small molecule ice recrystallization inhibitors. RSC Adv 5(28):21728–21732
32. Khan MS, Bavoh CB, Partoon B, Lal B, Bustam MA, Shariff AM (2017) Thermodynamic effect of ammonium based ionic liquids on CO_2 hydrates phase boundary. J Mol Liq 238(July):533–539
33. Bavoh CB, Khan MS, Lal B, Bt Abdul Ghaniri NI, Sabil KM (2018) New methane hydrate phase boundary data in the presence of aqueous amino acids. Fluid Phase Equilib 478:129–133
34. Brustad S, Løken KP, Waalmann JG (2005) Hydrate prevention using MEG instead of MeOH: impact of experience from major Norwegian developments on technology selection for injection and recovery of MEG. In: Offshore technology conference, no. OTC 17355, p 10
35. Khan MS, Lal B, Keong LK, Ahmed I (2019) Tetramethyl ammonium chloride as dual functional inhibitor for methane and carbon dioxide hydrates. Fuel 236:251–263
36. Sa J-H et al (2016) Inhibition of methane and natural gas hydrate formation by altering the structure of water with amino acids. Sci Rep 6:1–9
37. Bavoh CB, Partoon B, Lal B, Kok Keong L (2017) Methane hydrate-liquid-vapour-equilibrium phase condition measurements in the presence of natural amino acids. J Nat Gas Sci Eng 37:425–434
38. Park J, Lee H, Seo Y, Tian W, Wood CD (2016) Performance of polymer hydrogels Incorporating thermodynamic and kinetic hydrate inhibitors. Energy Fuels 30(4):2741–2750
39. Avula VR, Gardas RL, Sangwai JS (2015) An efficient model for the prediction of CO2 hydrate phase stability conditions in the presence of inhibitors and their mixtures. J Chem Thermodyn 85:163–170
40. Del Villano L, Kelland MA (2010) An investigation into the kinetic hydrate inhibitor properties of two imidazolium-based ionic liquids on Structure II gas hydrate. Chem Eng Sci 65(19):5366–5372
41. Babu P, Yao M, Datta S, Kumar R, Linga P (2014) Thermodynamic and kinetic verification of tetra-n-butyl ammonium nitrate ($TBANO_3$) as a promoter for the clathrate process applicable to precombustion carbon dioxide capture. Environ Sci Technol 48(6):3550–3558
42. Tariq M, Connor E, Thompson J, Khraisheh M, Atilhan M, Rooney D (2016) Doubly dual nature of ammonium-based ionic liquids for methane hydrates probed by rocking-rig assembly. RSC Adv 6(28):23827–23836
43. Smiglak M et al (2014) Ionic liquids for energy, materials, and medicine. Chem Commun 50(66)
44. Kim KS, Kang JW, Kang SP (2011) Tuning ionic liquids for hydrate inhibition. Chem Commun 47(22):6341–6343
45. Zare M, Haghtalab A, Ahmadi AN, Nazari K (2013) Experiment and thermodynamic modeling of methane hydrate equilibria in the presence of aqueous imidazolium-based ionic liquid solutions using electrolyte cubic square well equation of state. Fluid Phase Equilib 341:61–69

46. Keshavarz L, Javanmardi J, Eslamimanesh A, Mohammadi AH (2013) Experimental measurement and thermodynamic modeling of methane hydrate dissociation conditions in the presence of aqueous solution of ionic liquid. Fluid Phase Equilib 354(02):312–318
47. Sun SC, Liu CL, Meng QG (2015) Hydrate phase equilibrium of binary guest-mixtures containing CO2 and N2 in various systems. J Chem Thermodyn 84:1–6
48. Bavoh CB et al (2016) COSMO-RS: an ionic liquid prescreening tool for gas hydrate mitigation. Chinese J Chem Eng 24(11):1619–1624
49. Tariq M, Rooney D, Othman E, Aparicio S, Atilhan M, Khraisheh M (2014) Gas hydrate inhibition: a review of the role of ionic liquids. Ind Eng Chem Res 53(46):17855–17868
50. Qasim A, Khan MS, Lal B, Shariff AM (2019) Phase equilibrium measurement and modeling approach to quaternary ammonium salts with and without monoethylene glycol for carbon dioxide hydrates. J Mol Liq 282:106–114
51. Cha JH, Ha C, Kang SP, Kang JW, Kim KS (2016) Thermodynamic inhibition of CO2 hydrate in the presence of morpholinium and piperidinium ionic liquids. Fluid Phase Equilib 413:75–79
52. Israelachvili JN (2011) Intermolecular and surface forces, 3rd edn. Elsevier, Waltham
53. Jonathan CE, Taylor TK (2004) Advances in the study of gas hydrates, 1st edn. Kluwer Academic Publishers, New york, Boston, Dordrecht, London, Moscow
54. Perrin A, Musa OM, Steed JW (2013) The chemistry of low dosage clathrate hydrate inhibitors. Chem Soc Rev 42(5):1996–2015
55. Qasim A, Khan MS, Lal B, Ismail MC, Rostani K (2020) Quaternary ammonium salts as thermodynamic hydrate inhibitors in the presence and absence of monoethylene glycol for methane hydrates. Fuel 259, no. July 2019, p 116219
56. Cha M et al (2013) Thermodynamic and kinetic hydrate inhibition performance of aqueous ethylene glycol solutions for natural gas. Chem Eng Sci 99:184–190
57. Yang J, Chapoy A, Mazloum S, Tohid B (2011) Development of a hydrate inhibition monitoring system by integration of acoustic velocity and electrical conductivity measurements
58. Heng-Joo Ng DBR, Ng HJ, Robinson DB (1985) Hydrate formation in systems containing methane, ethane, propane, carbon dioxide or hydrogen sulfide in the presence of methanol. Fluid Phase Equilib 21, no. Elsevier Science, pp 145–155
59. Anderson FE, Prausnitz JM (1986) Inhibition of gas hydrates by methanol. AIChE J 32(8):1321–1333
60. Bavoh CB, Khan MS, Ting VJ (2018) The effect of acidic gases and thermodynamic inhibitors on the hydrates phase boundary of synthetic Malaysia natural gas the effect of acidic gases and thermodynamic inhibitors on the hydrates phase boundary of synthetic Malaysia natural gas. In: IOP Conference series: materials science and engineering paper, pp 1–10
61. Partoon B, Wong NMS, Sabil KM, Nasrifar K, Ahmad MR (2013) A study on thermodynamics effect of [EMIM]-Cl and [OH-C2MIM]-Cl on methane hydrate equilibrium line. Fluid Phase Equilib 337:26–31
62. Khan I, Kurnia KA, Sintra TE, Saraiva JA, Pinho SP, Coutinho JAP (2014) Assessing the activity coefficients of water in cholinium-based ionic liquids: Experimental measurements and COSMO-RS modeling. Fluid Phase Equilib 361:16–22
63. Gupta P, Chandrasekharan Nair V, Sangwai JS (2018) Phase equilibrium of methane hydrate in the presence of aqueous solutions of quaternary ammonium salts. J Chem Eng Data 63(7):2410–2419
64. Kang SP, Shin JY, Lim JS, Lee S (2014) Experimental measurement of the induction time of natural gas Hydrate and its prediction with polymeric kinetic inhibitor. Chem Eng Sci 116:817–823
65. Khan MS, Cornelius BB, Lal B, Bustam MA (2018) Kinetic assessment of tetramethyl ammonium hydroxide (ionic liquid) for carbon carbon dioxide, methane and binary mix gas hydrates. In: Recent advances in ionic liquids, no. September 2018, pp 159–179
66. Del Villano L, Kommedal R, Kelland MA (2008) Class of kinetic hydrate inhibitors with good biodegradability. Energy Fuels 22(5):3143–3149
67. Yin Z, Chong ZR, Tan HK, Linga P (2016) Review of gas hydrate dissociation kinetic models for energy recovery. J Nat Gas Sci Eng 35:1362–1387

68. Kelland MA (2006) History of the development of low dosage hydrate inhibitors. Energy Fuels 20(3):825–847
69. Roosta H, Dashti A, Mazloumi SH, Varaminian F (2016) Inhibition properties of new amino acids for prevention of hydrate formation in carbon dioxide-water system: experimental and modeling investigations. J Mol Liq 215:656–663
70. Carver TJ, Drew MGB, Rodger PM (1995) Inhibition of crystal-growth in methane hydrate. J Chem Soc Trans 91(19):3449–3460
71. Nashed O, Sabil KM, Ismail L, Japper-Jaafar A, Lal B (2017) Mean induction time and isothermal kinetic analysis of methane hydrate formation in water and imidazolium based ionic liquid solutions. J Chem Thermodyn, pp. 1–8
72. Khan MS, Bavoh CB, Partoon B, Nashed O, Lal B, Mellon NB (2018) Impacts of ammonium based ionic liquids alkyl chain on thermodynamic hydrate inhibition for carbon dioxide rich binary gas. J Mol Liq 261:283–290
73. Xu Y, Yang M, Yang X (2010) Chitosan as green kinetic inhibitors for gas hydrate formation. J Nat Gas Chem 19(4):431–435
74. Erfani A, Varaminian F, Muhammadi M (2013) Gas hydrate formation inhibition using low dosage hydrate inhibitors. In: 2nd National Iranian conference on gas hydrate (NICGH)
75. Talaghat MR (2014) Enhancement of the performance of modified starch as a kinetic hydrate inhibitor in the presence of polyoxides for simple gas hydrate formation in a flow mini-loop apparatus. J Nat Gas Sci Eng 18:7–12
76. Nguyen NN, Nguyen AV (2017) Hydrophobic effect on gas hydrate formation in the presence of additives. Energy & Fuels, p. acs.energyfuels.7b01467
77. Srivastava V, Chauhan DS, Joshi PG, Maruthapandian V, Sorour AA, Quraishi MA (2018) PEG-functionalized Chitosan: a biological macromolecule as a novel corrosion inhibitor. ChemistrySelect 3(7):1990–1998
78. Ke W, Kelland MA (2016) Kinetic hydrate inhibitor studies for gas hydrate systems: a review of experimental equipment and test methods. Energy Fuels 30:10015–10028
79. Perfeldt CM et al (2014) Inhibition of gas hydrate nucleation and growth: efficacy of an antifreeze protein from the longhorn beetle rhagium mordax. Energy Fuels 28:3666–3672
80. Qasim A, Khan MS, Lal B, Shariff AM (2019) A perspective on dual purpose gas hydrate and corrosion inhibitors for flow assurance. J Pet Sci Eng 183:106418
81. Park J, Kim H, Sheng Q, Wood CD, Seo Y (2017) Kinetic hydrate inhibition performance of poly(vinyl caprolactam) modified with corrosion inhibitor groups. Energy Fuels 31(9):9363–9373
82. Aman ZM, Koh CA (2016) Interfacial phenomena in gas hydrate systems. Chem Soc Rev 45:1678–1690
83. Koh CA, Sloan ED, Sum AK, Wu DT (2011) Fundamentals and applications of gas hydrates. Ann Rev Chem Biomolec Eng 2:237–257
84. Klomp UC, Kruka VC, Reijnhart R (1996) International Patent Application WO 95/17579, 1995
85. Huo Z, Freer E, Lamar M, Sannigrahi B, Knauss DM, Sloan ED (2001) Hydrate plug prevention by anti-agglomeration. Chem Eng Sci 56(17):4979–4991
86. Kelland MA, Svartås TM, Andersen LD (2009) Gas hydrate anti-agglomerant properties of polypropoxylates and some other demulsifiers. J Pet Sci Eng 64(1–4):1–10
87. Xiao C, Wibisono N, Adidharma H (2010) Dialkylimidazolium halide ionic liquids as dual function inhibitors for methane hydrate. Chem Eng Sci 65(10):3080–3087
88. Peng X, Hu Y, Liu Y, Jin C, Lin H (2010) Separation of ionic liquids from dilute aqueous solutions using the method based on CO_2 hydrates. J Nat Gas Chem 19(1):81–85
89. Khan MS, Lal B, Keong LK, Sabil KM (2018) Experimental evaluation and thermodynamic modelling of AILs alkyl chain elongation on methane riched gas hydrate system. Fluid Phase Equilib 473:300–309

Chapter 3
Application of Ionic Liquids in Corrosion Inhibition (CI)

3.1 Introduction

Corrosion is characterized as the degradation of a material due to electrochemical reactions, specifically for metallic materials (e.g. within pipelines, the formation of solid iron carbonate, $FeCO_3$ [1, 2], is one such cause of corrosion).

The flowline is deteriorated mainly due to an erosion and fretting corrosion. At high flowrate within the oil and gas pipeline, it becomes more vulnerable to corrosion issues. The surface load with high-speed movement is mostly responsible for causing fretting corrosion [3]. Pipeline suffers from erosion corrosion when it carries slurries and small particulate liquids inside it. As rough surfaces often come into contact with each other during transport, this produces high levels of friction within the system between the flow content and pipeline. The surface protective coating is destroyed through the friction build-up, and the metal is unprotected where it can be further corroded by the surroundings. The losses suffered by the industry from erosion corrosion contribute to a significant loss in terms of material degradation [4–6].

For flow control purposes, a suitable corrosion suppressant is used to correctly treat corrosion-related problems. The suitability of a corrosion suppressor depends upon its solubility in various acidic, basic and neutral solutions along with the ability to endure harsh operational conditions and environment. Chemicals for this application are classified under neutralizers, scavengers and absorbers, or film formers. In refineries, corrosion is commonly caused by hydrochloric acid and sulphuric acids. The concentration of these acidic chemicals is low in the process stream it can get high in condensate streams of heat exchangers or distillation column. The use of neutralizers reduces the concentration of hydrogen ions, effectively preventing corrosive conditions. Alkyl amines, ammonia and sodium hydroxide are some of the more widely used neutralizing compounds [7, 8].

For an inhibitor to be considered appropriate for use within a specific system, its physical conditions must be factored in. The condensation profiles of inhibiting compounds and acids should correlate, so that they will be present in the system at

© The Author(s), under exclusive licence to Springer Nature Switzerland AG 2021   33
B. Lal et al., *Ionic Liquids in Flow Assurance*,
SpringerBriefs in Petroleum Geoscience & Engineering,
https://doi.org/10.1007/978-3-030-63753-8_3

all periods of the corrosion formation [2]. The use of ammonia is considered as a cost-effective solution, but it is difficult to control its condensate stream solubility as it evaporates readily, rendering its use to have low efficacy [9, 10].

Scavengers are used to counter corrosion in flow assurance strategies. This involves the use of a removal system for 'scavenging' dissolved contaminants that can cause corrosion. For example, a typical steam stripping system can still leave behind small amounts of oxygen that may contribute to corrosive problems. Because of this, synthetic scavenging inhibitors are mixed into the system as an economically feasible additive for contaminant removal [11].

Similar to neutralizing and scavenging methods, it is also economically beneficial to inject compounds forming films within the tube which act as inhibitors of corrosion. To resist corrosion, these chemicals serve as a barrier. Due to strong interactions, the corrosion suppressing compounds form a protective film on the surface of the metal. These chemicals form protective layer through electrostatic adsorption or in some cases through chemisorption thereby significantly undermining corrosion effect [12, 13].

When the corrosion suppressing agent adsorbs to the steel surface, it covers the surface by forming the protective layer, slowing down the electrochemical reaction. The level of metal protection is proportional to the fraction of the surface that the suppressing compound is covering. Surface coverage (θ) and inhibition efficiency are some of the key measures used to assess the efficacy of corrosion inhibitors [13, 14].

3.2 Essential Concepts of Corrosion Formation

Besides the issue of gas hydrates, corrosion formation is another problematic issue affecting the pipeline flow assurance integrity [15–18]. The occurrence of carbon dioxide and hydrogen sulphide which have acidic nature forms acid gases so the pipeline becomes vulnerable to corrosion and hydrate development inside it thereby compromising normal operation. In this chapter, therefore, types of corrosion formation and factors affecting it in flow assurance industry have been discussed.

3.2.1 Corrosion Formation

Degradation of metals in aqueous environments is the basic principle of corrosion. The corrosion reaction on the metal surface is classified as reduction-oxidation (redox) electrochemical reaction [3, 19]. The metal loses electrons as it dissolutes, and electrons are transferred to its surface. Oxygenated water is decreased due to this electrochemical effect and the continual occurrence of the process results in eventual failure of the pipeline [3].

3.2.2 Corrosion Mechanism

To prevent corrosion, the process of corrosion needs to be understood. The electrochemical corrosion reaction is mentioned as redox reaction in Eqs. (3.1–3.6) [2, 3]. The first reaction occurs at the anode which is the metal oxidation reaction given in Eq. (3.2). In a corrosive environment, metal dissolutes and cations are released during this process. The released cations work as electrolytes. In addition, free electrons are produced that pass through the metal sheet. By comparison, these free electrons amass at the metal surface and cause a potential difference which is neutralized at the cathode as hydrogen gas is formed by hydrogen ions as written in Eq. (3.3). The presence of electrolyte is necessary and plays a major role in transport of ions on the surface along with cathodic and anodic sites. Therefore, it is essential to have electrolyte solution in order to accomplish cathode and anode circuit as well as the free electrons transport at the metal surface. The oxidation process produces anodic current which in turn reduces oxygenated water in a neutral environment. Hydroxyl ions are formed by this reduction process. The reaction is given in Eq. 3.4. A connecting medium is materialized between anode and cathode in the presence of electrolyte. The dissolved cations and anions form insoluble deposits of iron hydroxides as given in Eq. (3.5) [20, 21]. Chemicals that are involved in the anodic and cathodic reactions typically are part of the mixed inhibitors.

Equation (3.1) shows the overall electrochemical corrosion reaction:

$$Fe + 2H^+ \rightarrow Fe^{+2} + H_2 \tag{3.1}$$

Oxidation half-reaction can be given as according to Eq. (3.2):

$$Fe \rightarrow Fe^{+2} + 2e^- \tag{3.2}$$

The reduction reaction is mentioned in Eq. (3.3):

$$2H^+ + 2e^- \rightarrow H_2 \tag{3.3}$$

In the presence of oxygen contaminant, Eqs. (3.4) and (3.5) show the chemical reaction:

$$O_2 + 2H_2O + 4e^- \rightarrow 4OH^- \tag{3.4}$$

$$Fe^{+2} + 2OH^- \rightarrow Fe(OH)_2 \tag{3.5}$$

Ferrous carbonate, $FeCO_3$, is the corrosion-causing agent which consists of solid form and is the final product of the reaction. Equation (3.6) shows the formation of $FeCO_3$.

$$Fe + CO_2 + H_2O \rightarrow FeCO_3 + H_2 \tag{3.6}$$

The corrosion inside the pipeline is dependent on various factors affecting the process stream. A work by Brondel et al. [22] explained some commonly occurring causes of offshore corrosion. One of the most probable causes is due to contact with sea water or rain. In order to address this issue, a zinc coating is used which acts as a sacrificial anode. In addition, cathodic protection is often used when a device uses opposing currents to corrosion.

The effect of corrosion can be felt throughout the pipelines, even in hydrocarbon process streams. Due to the severe conditions in deep-water environment, the flowline is prone to suffer from corrosion thus harming the safe flow. In order to address this problem, some of the economical techniques used by flow assurance industry include the addition of corrosion inhibitors inside the pipeline or to perform coating of the surface. [12, 23].

The hypothesis behind protection against corrosion is due to shielding effect [24]. For steel pipelines, the corrosion inhibitor interrupts the electrochemical reaction by adsorbing at the surface [23]. The efficacy of the mitigation performance can be assessed by the amount of the inside pipeline surface covered by the corrosion inhibitor. This phenomenon can be explained mathematically by studying the relationship between corrosion inhibitor concentration and surface coverage. The relationship between these two quantities is referred to as adsorption isotherm. The effectiveness of a certain corrosion inhibitor is assessed by developing adsorption isotherm. Some adsorption isotherms used in regard include Langmuir, Freundlich or Frumkin isotherm among others [25].

3.3 Classification of Corrosion in Pipeline

The flowline can face damage by four different types of corrosion occurring in it. These types are classified as fretting, uniform, stress and pitting corrosion. Among all these types of corrosion, it is easier to identify uniform corrosion on the basis of visual inspection, whereas other types of corrosion are localized and not noticeable. Stress and corrosion cracking is also caused by hydrogen ion mediation [26]. All the groups of corrosion listed here are unique in their development and behaviour. Figure 3.1 shows different types of corrosion affecting the oil and gas pipeline.

3.3.1 Pitting and Crevice Corrosions

Pitting and crevice corrosions are identical types and mostly occur in immobile water. It may also materialize in the CO_2, chlorine or oxygen environment. These types of corrosion are confined to a smaller area of metal surface, and the extent of damage is localized. It creates the voids at the metal surface, varying from tiny cavities in diameter to fairly shallow cavities. Corrosion of pitting was regularly found in the pipelines for both sweet and sour areas. Pitting corrosion is hard to detect at the

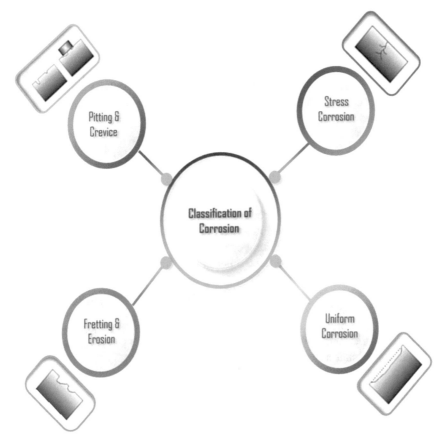

Fig. 3.1 Types of corrosion deteriorating pipeline surface

design point, as the cavities are developed and filled by corrosion materials over a continual period of time [26], potentially and unnoticeably leading to the formation of tiny narrow pits. However, this tiny pit can cause the entire pipeline system to collapse.

Crevice corrosion ensues due to metal coming in contact with the other metal surfaces or any other type of material which can contribute towards corrosion formation. Kennell and Evitts stated that crevice corrosion damages the same surface more as compared to the surrounding areas [27]. It is also more prevalent around protected parts such as gaskets, insulating material, threads, gaskets and joints.

3.3.2 Stress-Corrosion Cracking (SCC)

SCC damages the metal surface due to residual internal stresses. During the manufacturing stages, residual stresses are typically created by metal deformation. The deformation occurs due to process known as quenching which is caused by uneven transfer of heat during cooling. The quenching process damages the structure internally due to the volume shift [2].

3.3.3 Fretting and Erosion Corrosion

Fretting and erosion corrosion occurs due to high flow rate inside the oil and gas flowline. It may occur in the transportation process when the rough surfaces interact and suffer vibrations causing fretting corrosion [28, 29]. This is because the metal surfaces that are filled influence each other in their movement [30]. The scratching during the movement dissolves coating which protects metal surface, and the metal is continuously exposed towards corrosion effect. The metal degradation through erosion corrosion is known to be a major flow assurance problem in the oil and gas industries [4]. At offshore sites, pipeline can be damaged by fretting corrosion as it transports slurries and liquids containing particulate matter. Flow assurance industry spends a significant amount of capital to deal with the damage induced from this type of corrosion [31].

3.3.4 Uniform Corrosion

It is considered to be less damaging as the rate of corrosion can be predicted. It leaves homogeneous chemical composition and structure on metal surfaces [32]. Unlike localized corrosion which is hard to detect and mitigate, uniform type of corrosion deteriorates the metal surface uniformly and almost in an even manner. It is usually tackled by applying paints and coatings, or simply by specifying in the design stage for limited allowance of corrosion.

3.4 Corrosion Inhibitors

The offshore oil and gas industry faces the problem of corrosion as it deals with sea water and salts in harsh environments comprising of higher pressures and lower temperatures in the pipeline. Insertion of suitable corrosion inhibitors is employed to address this issue [11]. For this purpose, depending upon the suitability, scavenging, neutralizing or film-forming inhibitors are used [33–35]. Neutralizing inhibitors are added to the system to prevent corrosion inside the flowline by decreasing the concentration of the hydrogen ion in the system environment. Ammonia, alkyl amines and sodium hydroxide are widespread used neutralizers [35]. Neutralizer use typically requires physical conditions close to that of the pipeline and its contents. The neutralizers' condensation profile should match the acid's condensation profile so that it is available all the time in the system wherever there is a high potential for acid formation. Ammonia is also used as a cheap neutralizing agent, but it is not effective as it evaporates readily [36, 37]. Scavenger systems are implemented expressly to extract corrosive agents from oil and gas pipelines. Some companies usually set up a steam stripping device that doesn't eliminate the oxygen traces entirely, so synthetic scavenging inhibitors are implemented to further deoxygenate the stream. Film-forming corrosion inhibitors or barrier inhibitors function with an advantage as they do not interact with other corrosion-causing agents, e.g. acids. They create a protective barrier on the pipeline's metal surface through interactions involving electrostatic adsorption and chemisorption, to significantly reduce the penetration capabilities of corrosive substances [25, 38].

Independent of the categories above, corrosion inhibitors can be further classified as either an anodic, or cathodic, or mixed inhibitor. By testing the corrosion potential (E_{corr}) changes in the absence and presence of inhibitors, if the cathodic or anodic direction is found to be greater than 85 mV, then it can be categorized as cathodic or anodic type of inhibitor [3]. Unless the achieved value is found to be below 85 mV, then the inhibitor can be considered as an inhibitor of a mixed form. At the location of electron-donating atoms, anionic functional groups are adsorbed at the steel surface thus blocking the anodic sites. Similarly, cathodic sites are blocked by the cations. It was found that electrochemical conditions display a significant role in determining the appropriate type of corrosion inhibitor to be used for the corrosion inhibition application.

3.5 ILs as Corrosion Inhibitors

Ionic liquids composed of ammonium and imidazolium moieties have been used in corrosion mitigation studies along with thermodynamic and kinetic hydrate inhibition investigations [39, 40]. Furthermore, pyridazinium and pyrrolidinium groups have also been studied for their corrosion mitigation performance on the steel surface thereby protecting its integrity [41, 42]. The corrosion inhibition effect of ILs has been studied in all types of environments such as acidic, basic and neutral electrolytic media [43]. It includes the performance examination of ILs in HCl, NaCl, or H_2SO_4 solutions [44]. To protect mild steel surface corrosion in a salty environment of 3.5 wt% NaCl, Shamy et al. [44] used 1-butyl-1-methylpyrrolidinium trifluoromethylsulfonate, [BMPyrr]$^+$[Otf]$^-$ ionic liquid. The concentration of 3.5 wt% NaCl is the typical sea water salt concentration.

Investigators observed mixed-type corrosion inhibition. Ionic liquids, namely 1-butyl-2,3-dimethylimidazolium tetrafluoroborate [BDMIM]$^+$[BF$_4$]$^-$ and 1-ethyl-3-methylimidazolium tetrafluoroborate [EMIM]$^+$[BF$_4$]$^-$, were researched in corrosion inhibition application of mild steel in acidic conditions of 1 M HCl. Researchers found the mode of corrosion inhibition to be of mixed type for both ionic liquid chemicals [45]. The use of 3-(4-chlorobenzoylmethyl)-1-methylbenzimidazolium bromide [BMIMB]$^+$[Br]$^-$ by Kannan and co-workers revealed the mixed-type mitigation behaviour of this particular ionic liquid. It was used to protect mild carbon steel in an acidic media of 1 M HCl [46]. The mild steel corrosion protection by ionic liquid 1-butyl-3-methylimidazolium tetrachloroferrate, [BMIM]$^+$[FeCl$_4$],$^-$ in controlled and open environments was examined. Researchers used both weight loss and electrochemical techniques for the study [47]. Another ionic liquid, namely 1-butyl-3-methylimidazolium bromide [BMIM]$^+$[Br],$^-$ showed a shielding layer effect to protect the mild steel in an acidic media of 1 M HCl as studied by Ashassi et al. [48]. The results from both gravimetric and electrochemical methods showed mixed-type inhibition characteristics. The use of five imidazolium-based ILs including 1-ethyl-3-methylimidazolium acetate, [EMIM]$^+$[Ac]$^-$, 1-ethyl-3-methylimidazolium ethyl sulfate [EMIM]$^+$[EtSO$_4$]$^-$, 1-butyl-3-methylimidazolium acetate [BMIM]$^+$[Ac]$^-$, 1-butyl-3-methylimidazolium dicyanamide [BMIM]$^+$[DCA]$^-$ and 1-butyl-3-methylimidazolium thiocyanate [BMIM]$^+$[SCN]$^-$ showed their corrosion inhibiting properties for mild steel protection application. The experiments were done in an acidic environment of 1 M HCl by applying spectroscopic and electrochemical methods along with Monte Carlo simulations [40]. The experimental and simulation results showed mixed-type corrosion inhibition behaviour. In an aqueous monoethanolamine system, researchers examined the corrosion inhibition performances of ILs, namely 1-ethyl-methylimidazolium tetrafluoroborate [EMIM]$^+$[BF$_4$]$^-$ and 1-ethyl-3methylimidazolium trifluoromethanesulfonate [EMIM]$^+$[Otf]$^-$ [49]. The study was performed in an acidic surrounding containing 3% CO$_2$. Among these ionic liquids, [EMIM]$^+$[BF$_4$]$^-$ showed the best mitigation performance followed by [EMIM]$^+$[Otf]$^-$ and [BMIM]$^+$[Otf],$^-$ respectively. Yousefi et al. [43] determined the corrosion rates of six ILs using electrochemical method. All the considered ILs belonged to imidazolium-based cationic moiety. The experimental study was done in a corrosive environment of 2 M HCl. They revealed mixed-type corrosion mitigation behaviour. Murulana et al. [50] expanded the search for imidazolium-based ILs and used them in a 1 M HCl solution as a corrosion inhibitor for mild steel. The ILs comprised of butyl and propyl groups of 1-alkyl-3-methylimidazolium bis(trifluoromethyl-sulfonyl) imide [BMIM, PMIM]$^+$[NTf$_2$]$^-$, and propyl and hexyl groups of 1-alkyl-2,3-methylimidazolium bis(trifluoromethyl-sulfonyl) imide [PDMIM, HMIM]$^+$[NTf$_2$]$^-$. Corrosion rates were obtained for all the chemicals. Gravimetric and electrochemical analysis showed their mixed inhibition behaviour. It was found that [PDMIM]$^+$[NTf$_2$]$^-$ performed better than the other ionic liquids owing to propyl moiety in the structure [50]. The gravimetric and electrochemical corrosion inhibition results of two ionic liquids, i.e. 3-(3-phenylpropyl)-1-propyl-1H-imidazol-3-ium bromide (PPIB1) and 3-(4-phenoxybutyl)-1-propyl-1Himidazol-3-ium bromide (PPIB4) studied by Zarrouk et al., showed mixed-type corrosion suppression behaviour [51]. These ionic liquids were employed to protect carbon steel from deterioration in a corrosive environment of 1 M HCl.

Likewise, quaternary ammonium salts (QAS) have also been tested as suppressors of corrosion. QAS like ionic liquids also compromise of anionic and cationic moieties tightly held together. In QAS, the structure of the cation is quaternary. Using electrochemical impedance spectroscopy (EIS) and potentiodynamic polarization scans, the performance of cationic mixtures of cetyltrimethyl ammonium bromide (CTAB) and sodium dodecyl sulfate (SDS) in a 3.% NaCl electrolyte media was inspected. The experiments were performed for mild steel surface protection. Corrosion result analysis showed their mixed-type inhibition behaviour, while CTAB showed enhanced performance than SDS. CTAB is considered to be a cationic surfactant, while SDS has anionic properties. Researchers stated that CTAB showed better mitigation effect owing to a strong association between steel and surfactant surface polar head groups [52]. In addition, Chong et al. [53] used electrochemical and weight loss methods to examine the corrosion levels of organic salt containing a protic imidazolium cation and a 4-hydroxycinnamate anion in aqueous 0.01 M NaCl solution. The corrosion behaviours of 1-(4-sulfobutyl)-3-methylimidazolium hydrogen sulfate $[BsMIM]^+[HSO_4]^-$ and 1-(4-sulfobutyl)-3-methylimidazolium tetrafluoroborate $[BsMIM]^+[BF_4]$ were studied by Ma et al. [18] via EIS and polarization method in 1 M H_2SO_4 solution. The use of di-quaternary ammonium salts (QAS) as corrosion inhibitors to protect pipeline steel (API X65) in a corrosive environment of 1 M HCl was investigated by Hegazy et al. [14]. They synthesized and used three QAS including N-3-2-isopropyl dimethylammonio acetoxy propyl-N,N-dimethyldodecan-1-aminium chloride bromide, N-3-2-2-hydroxyethyl dimethylammonio acetoxy propyl-N,N-dimethyldodecan-1-aminium chloride bromide and N-3-2-phenyldiethylammonio acetoxy propyl- N,N-dimethyldodecan-1-aminium chloride bromide for the application. Gravimetric and electrochemical methods were used to examine the mitigation behaviour. All techniques showed the corrosion inhibition effect of the compounds. The polarization curves showed that the inhibition characteristics of the QASs were mixed in nature. The Langmuir isotherm was identified as a mechanism for physical adsorption [14].

The corrosion inhibition performance of $[C_4C_1IM]^+[FeCl_4]^-$ on A36 mild steel was examined in open and controlled environments using electrochemical and immersion techniques. The results acquired from both the techniques showed agreement. In an open environment, the corrosion rate found on the metal surface was found to be higher [47].

For the case of carbon steel 1020 corrosion, Rehman et al. assessed the performance of some room-temperature ionic liquids (RTILs). The chemicals employed for the investigation included 1-ethyl-3-methylimidazolium dicyanamide [EMIM]$^+$[DCA]$^-$ and 1-ethyl-3-methylimidazolium acetate [EMIM]$^+$[Ac]$^-$. The corrosion rate was found to change with differing concentrations. Among the used ionic liquids, [EMIM]$^+$[Ac]$^-$ exhibited the best relative performance and showed a corrosion inhibition efficiency of about 88% [54].

Using the potentiodynamic polarization process, Koswari et al. used tetra-n-butyl ammonium methioninate to mitigate corrosion in an acidic environment. The study was performed to protect mild steel surface in 1.0 M HCl solution. Potentiodynamic polarization exposed methioninate of tetra-n-butylammonium as a mixed-type corrosion inhibitor with predominantly anodic action. According to Freundlich isotherm, the adsorption of tetra-n-butyl ammonium methioninate was found on the soil. It has been concluded that physisorption followed the adsorption process [55].

Due to the presence of both anionic and cationic parts, mostly ionic liquids act as mixed-type corrosion inhibitors as both electropositive and electronegative moieties participate in the application. The inhibition mechanism of ionic liquids for protecting the pipeline surface is that these compounds form a protective layer at the steel surface achieved either through physisorption or chemisorption. Primarily, the surface adsorption approaches Langmuir, Freundlich or Frumkin isotherm among other acceptable isotherms. The molecular mechanism is explained using the functional theory of density (DFT), as defined by Umoren et al. [56]. The adsorption phenomenon is observed due to donor–acceptor interactions between the unoccupied metal surface d-orbital and valence electrons of ionic liquid. Hydrophobic behaviour increases with the increase in alkyl chain length thus inducing the tendency of corrosion inhibitors to adsorb on the surface. In this way, organic compounds including ionic liquids are able to act as film-forming corrosion inhibitors. Hence, the influencing factors affecting corrosion mitigation behaviour of chemicals include pH of the solution, hydrodynamic conditions, and alkyl chain length of corrosion suppressor. The value of pH is considered important as it should be towards neutral side to achieve better results in terms of corrosion inhibition. Regarding chain length, longer alkyl chain length will produce a better shielding effect on the extent of adsorption onto the pipeline steel surface. Figure 3.2 shows the factors which play an important role in identifying the system's corrosion behaviour.

Table 3.1 shows the performance of some ILs as corrosion inhibitors, studied for their potential application in oil and gas flow pipelines. The corrosion inhibition performance of ILs has been investigated in preceding literature, and the quantification of corrosion suppression in terms of inhibition efficiency percentage (IE%) is stated in Table 3.1. In terms of corrosion inhibition efficiency, performance of [EMIM]$^+$[EtSO$_4$]$^-$ is found to be better than other compounds.

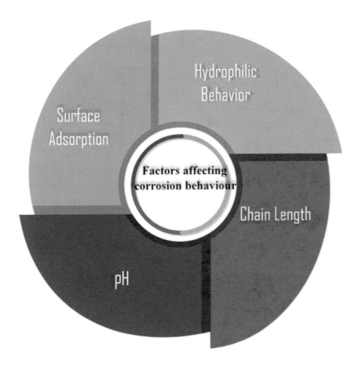

Fig. 3.2 Salient features involved in corrosion inhibition performance

Table 3.1 Corrosion inhibition performance of Ionic liquids (ILs)

Ionic liquids	Structure	Inhibitor specification		Results	References
		Concentration	Electrolytic media (25 °C)		
[EMIM]⁺[Cl]⁻		1, 5 mM	3.5% NaCl	At 5 mM, inhibition efficiency is found to be 70%	[57]
[EMIM]⁺[BF₄]⁻		50–500 ppm	1 M HCl	At 500 ppm inhibition efficiency is found to be 77.93%	[45]
[BMIM]⁺[Br]⁻		2–20 mM	1 M HCl	At 20 mM, inhibition efficiency is reported to be 94%	[48]
[EMIM]⁺[EtSO₄]⁻		100, 300, 500 ppm	1 M HCl	At 500 ppm inhibition efficiency of 92.75% is determined	[40]
[BMIM]⁺[Cl]⁻		0.05–3 wt%	2 M HCl	At 1 wt%, it shows the inhibition efficiency of 79%	[43]

(continued)

Table 3.1 (continued)

Ionic liquids	Structure	Inhibitor specification		Results	References
		Concentration	Electrolytic media (25 °C)		
[BMIM]$^+$[PF$_6$]$^-$		0.1–0.5 wt%	2 M HCl	At 0.1 wt%, the IE is determined to be 86%	[43]
[C$_4$C$_1$IM][FeCl$_4$]		1 wt%	–	Corrosion rate of mild steel for the IL in open and control environment is found to be 1.42 mm/yr and 25.1 mm/yr, respectively	[47]
[EMIM][Ac] (in 30 wt% MEA aqueous solution)		–	CO$_2$ Capture System	Researchers used electrochemical method and the CR was found to be 10.74 mm/yr. EMIMAc showed better results than EMIMCCA	[54]
[EMIM][DCA] (in 30 wt% MEA aqueous solution)		–	CO$_2$ Capture System	Electrochemical method was used and CR of 2.29 mm/yr was obtained	[54]
[TBA][L-Met]		0.0000133–0.00159 M	1 M HCl	Electrochemical method was used to find the corrosion performance. The adsorption showed Freundlich isotherm. Fit	[55]

3.6 Summary

This chapter discusses the role of ionic liquids as corrosion inhibitors as potential candidates for usage in oil and gas industry pipelines. The researchers have investigated different types of ionic liquids in corrosion mitigation applications. The corrosion mitigation performance of most ionic liquids has mainly been in acidic environments of electrolytic media such as HCl or H_2SO_4. It is found from the literature that the IL chemical 1-ethyl-3-methylimidazolium ethyl sulfate $[EMIM]^+[EtSO_4]^-$ achieved better results for corrosion inhibition in comparison with the other chemicals within the scope of analysis. The key factor in corrosion inhibition is the ability of ILs to adsorb on the surface of the steel thus forming a protective layer. This formation of protective coating on metal surface protects it from further corrosion. Along with it, optimal chain length and pH of the system are also of importance. Thus, it is concluded that many ionic liquids show potential in commercial usage of corrosion suppression and the research can be enhanced further.

References

1. Mendonca GLF, Costa SN, Freire VN, Casciano PNS, Correia AN, de Lima-Neto P (2017) Understanding the corrosion inhibition of carbon steel and copper in sulphuric acid medium by amino acids using electrochemical techniques allied to molecular modelling methods. Corros Sci 115:41–55
2. Saji VS, Umoren SA (2020) Corrosion inhibitors in the oil and gas industry, 1st edn. Wiley-VCH
3. Tiu BDB, Advincula RC (2015) Polymeric corrosion inhibitors for the oil and gas industry: design principles and mechanism. React Funct Polym 95:25–45
4. Rajahram SS, Harvey TJ, Wood RJK (2009) Erosion-corrosion resistance of engineering materials in various test conditions. Wear 267(1–4):244–254
5. Theyab MA (2018) Fluid flow assurance Issues: literature review. SciFed J Pet 2(1):1–11
6. Obanijesu EO, Gubner R, Barifcani A, Pareek V, Tade MO (2014) The influence of corrosion inhibitors on hydrate formation temperature along the subsea natural gas pipelines. J Pet Sci Eng 120:239–252
7. Menendez CM et al (2014) New sour gas corrosion inhibitor compatible with kinetic hydrate inhibitor. In: International petroleum technology conference, pp 1–9
8. Verma C, Ebenso EE, Quraishi MA (2018) Ionic liquids as green corrosion inhibitors for industrial metals and alloys. Green Chem I:103–132
9. Obanijesu EO (2012) Corrosion and hydrate formation in natural gas pipelines. Curtin University, Australia
10. Obanijesu EO, Pareek V, Tade MO (2010) Hydrate formation and its influence on natural gas pipeline internal corrosion rate. Environment 62(5–6):164–173
11. Burgazli CR, Navarrete RC, Mead SL (2005) New dual purpose chemistry for gas hydrate and corrosion inhibition. J Can Pet Technol 44(11):47–50
12. Schweinsberg DP, Ashwortht V, Ashworth V, Ashwortht V (1988) The inhibition of the corrosion of pure iron in 0.5 M sulphuric acid by n-alkyl quaternary ammonium iodides. Corros Sci 28(6):539–545
13. Israelachvili JN (2011) Intermolecular and surface forces, 3rd edn. Elsevier, Waltham
14. Hegazy MA, Abdallah M, Awad MK, Rezk M (2014) Three novel di-quaternary ammonium salts as corrosion inhibitors for API X65 steel pipeline in acidic solution. Part I: experimental results. Corros Sci 81:54–64

15. Sivabalan V, Hassan NA, Qasim A, Lal B, Bustam MA (2020) Density measurement of aqueous tetraethylammonium bromide and tetraethylammonium iodide solutions at different temperatures and concentrations. South African J Chem Eng 32:62–67

16. Kermani MB, Morshed A (2003) Carbon dioxide corrosion in oil and gas production—A compendium. Crit Rev Corros Sci Eng 59(8):659–683

17. Khan MS, Lal B, Keong LK, Ahmed I (2019) Tetramethyl ammonium chloride as dual functional inhibitor for methane and carbon dioxide hydrates. Fuel 236, no. May 2018, pp 251–263

18. Ma Y, Han F, Li Z, Xia C (2016) Acidic-functionalized ionic liquid as corrosion inhibitor for 304 stainless steel in aqueous sulfuric acid. ACS Sustain Chem Eng 4(9):5046–5050

19. Glasstone S, Lewis D (1960) Elements of physical chemistry, 2nd ed. The Macmillan Company of India Limited

20. Hammami A, Ratulowski J (2007) Precipitation and deposition of asphaltenes in production systems: a flow assurance overview. Asph Heavy Oils, Pet, pp 617–660

21. Verma C, Ebenso EE, Quraishi MA (2017) Ionic liquids as green and sustainable corrosion inhibitors for metals and alloys: an overview. J Mol Liq 233(2016):403–414

22. Brondel D, Edwards R, Hayman A, Hill D, Mehta S, Semerad T (1994) Corrosion in the oil industry. Oilf Rev 6(4):4–18

23. Raja PB et al (2016) Reviews on corrosion Inhibitors: a short view. Chem Eng Commun 203(9):1145–1156

24. Hedges B et al (2006) Paper No. 06120. *Corrosion*, no. 06120, pp 1–30

25. Nešić S (2007) Key issues related to modelling of internal corrosion of oil and gas pipelines - A review. Corros Sci 49(12):4308–4338

26. Heppner KL, Evitts RW (2008) Modelling of the effect of hydrogen ion reduction on the crevice corrosion of titanium. Environ Crack Mater 1:95–104

27. Kennell GF, Evitts RW (2009) Crevice corrosion cathodic reactions and crevice scaling laws. Electrochim Acta 54(20):4696–4703

28. Srinivasan S, Kane RD (2003) Critical issues in the application and evaluation of a corrosion prediction model for oil and gas systems. *Corros*, no 03640, pp 1–18

29. Xiao K, gang Dong C, gang LI X, F. ming Wang (2008) Corrosion products and formation mechanism during initial stage of atmospheric corrosion of carbon steel. J Iron Steel Res Int 15(5):42–48

30. Srinivasan S, Kane RD (2003) Critical issues in the application and evaluation of a corrosion prediction model for oil and gas systems. NACE - Int Corros Conf Ser 2003(03640):1–18

31. Wang F, Wang LF (1998) Polydimethylsiloxane modification of segmented thermoplastic polyurethanes and polyureas polydimethylsiloxane modification of segmented thermoplastic polyurethanes and polyureas. Virginia Polytechnic Institute and State University

32. Nordsveen M, Nešić S, Nyborg R, Stangeland A (2003) A mechanistic model for carbon dioxide corrosion of mild steel in the presence of protective iron carbonate films—Part 1: theory and verification. Corrosion 59(5):443–456

33. Sivabalan V, Walid B, Madec Y, Qasim A, Lal B (2020) Corrosion inhibition study on glycerol as simultaneous gas hydrate and corrosion inhibitor in gas pipelines. Malaysian J Anal Sci 24(1):62–69

34. Laamari R, Villemin D (2011) Corrosion inhibition of carbon steel in hydrochloric acid 0.5 M by hexa methylene diamine tetramethyl-phosphonic acid. Arab J Chem 4(3):271–277

35. Liu Z, Jackson TS, Ramachandran S (2018) Synergic corrosion inhibitors. US2018/0201826A1

36. Kelland MA (2014) Production chemicals for the oil and gas industry, 2nd edn. CRC Press, New York

37. Parkash S (2003) Refining process handbook, 1st edn. Gulf Professional Publishing is an imprint of Elsevier, Amsterdam; Boston; Heidelberg; London; New York; Oxford; Paris; San Diego; San Francisco; Singapore; Sydney; Tokyo

38. Umoren SA, Obot IB (2008) Polyvinylpyrrolidone and polyacrylamide as corrosion inhibitors for mild steel in acidic medium. Surf Rev Lett 15(3):277–286

39. Pisarova L, Gabler C, Dörr N, Pittenauer E, Allmaier G (2012) Thermo-oxidative stability and corrosion properties of ammonium based ionic liquids. Tribol Int 46(1):73–83
40. Yesudass S, Olasunkanmi LO, Bahadur I, Kabanda MM, Obot IB, Ebenso EE (2016) Experimental and theoretical studies on some selected ionic liquids with different cations/anions as corrosion inhibitors for mild steel in acidic medium. J Taiwan Inst Chem Eng 64:252–268
41. Bousskri A et al (2015) Corrosion inhibition of carbon steel in aggressive acidic media with 1-(2-(4-chlorophenyl)-2-oxoethyl)pyridazinium bromide. J Mol Liq 211:1000–1008
42. Tawfik SM (2016) Ionic liquids based gemini cationic surfactants as corrosion inhibitors for carbon steel in hydrochloric acid solution. J Mol Liq 216:624–635
43. Yousefi A, Javadian S, Dalir N, Kakemam J, Akbari J (2015) Imidazolium-based ionic liquids as modulators of corrosion inhibition of SDS on mild steel in hydrochloric acid solutions: experimental and theoretical studies. RSC Adv 5:11697–11713
44. El-Shamy AM, Zakaria K, Abbas MA, Zein El Abedin S (2015) Anti-bacterial and anti-corrosion effects of the ionic liquid 1-butyl-1-methylpyrrolidinium trifluoromethylsulfonate. J Mol Liq 211:363–369
45. Sasikumar Y et al (2015) Experimental, quantum chemical and Monte Carlo simulation studies on the corrosion inhibition of some alkyl imidazolium ionic liquids containing tetrafluoroborate anion on mild steel in acidic medium. J Mol Liq 211:105–118
46. Kannan P, Karthikeyan J, Murugan P, Rao TS, Rajendran N (2016) Corrosion Inhibition effect of novel methyl benzimidazolium Ionic liquid for carbon steel in HCl medium. J Mol Liq 221:368–380
47. Ullah S, Bustam MA, Shariff AM, Gonfa G, Izzat K (2016) Experimental and quantum study of corrosion of A36 mild steel towards 1-butyl-3-methylimidazolium tetrachloroferrate ionic liquid. Appl Surf Sci 365:76–83
48. Ashassi-Sorkhabi H, Es'haghi M (2009) Corrosion inhibition of mild steel in acidic media by [BMIm]Br Ionic liquid. Mater Chem Phys 114(1):267–271
49. Acidi A, Hasib-ur-Rahman M, Larachi F, Abbaci A (2014) Ionic liquids [EMIM][BF4], [EMIM][Otf] and [BMIM][Otf] as corrosion inhibitors for CO2 capture applications. Korean J Chem Eng 31(6):1043–1048
50. Murulana LC, Singh AK, Shukla SK, Kabanda MM, Ebenso EE (2012) Experimental and quantum chemical studies of some bis(trifluoromethyl- sulfonyl) imide imidazolium-based ionic liquids as corrosion inhibitors for mild steel in hydrochloric acid solution. Ind Eng Chem Res 51(40):13282–13299
51. Zarrouk A et al (2012) Some new ionic liquids derivatives: synthesis, characterization and comparative study towards corrosion of C-steel in acidic media. J Chem Pharm Res 4(7):3427–3436
52. Javadian S, Yousefi A, Neshati J (2013) Synergistic effect of mixed cationic and anionic surfactants on the corrosion inhibitor behaviour of mild steel in 3.5% NaCl. Appl Surf Sci 285, no. PARTB, pp 674–681
53. Chong AL, Mardel JI, MacFarlane DR, Forsyth M, Somers AE (2016) Synergistic corrosion inhibition of mild steel in aqueous chloride solutions by an imidazolinium carboxylate salt. ACS Sustain Chem Eng 4(3):1746–1755
54. Hasib-Ur-Rahman M, Larachi F (2013) Prospects of using room-temperature ionic liquids as corrosion inhibitors in aqueous ethanolamine-based CO2 capture solvents. Ind Eng Chem Res 52(49):17682–17685
55. Kowsari E et al (2016) In situ synthesis, electrochemical and quantum chemical analysis of an amino acid-derived ionic liquid inhibitor for corrosion protection of mild steel in 1 M HCl solution. Corros Sci 112:73–85
56. Umoren SA, Eduok UM (2016) Application of carbohydrate polymers as corrosion inhibitors for metal substrates in different media: a review. Carbohydr Polym 140:314–341
57. Sherif ESM, Abdo HS, El Abedin SZ (2015) Corrosion inhibition of cast iron in arabian gulf seawater by two different ionic liquids. Materials (Basel) 8(7):3883–3895

Chapter 4
Application of Ionic Liquids in Wax, Scale and Asphaltene Deposition Control

4.1 Introduction

4.1.1 Wax Control in Flow Assurance

Commonly in the oil and gas industry, wax is identified as a type of solid matter that is dissolved during heating or cooling procedures with several key observable measures [1, 2]. One of such is the wax appearance temperature (WAT) and can be explained as the moment where the first crystals of wax start appearing [3, 4]. The instant temperature at which these crystal structures start dissolving back in the oil is known as wax dissolution temperature (WDT) [1, 5]. In terms of magnitude, the temperature of WDT is usually greater than WAT. Wax deposition, which occurs while the oil is being transported or produced, is a serious issue and considered a production constraint. It is constituted by n-paraffins and comprises of linear chains. In addition to this, it has small amounts of branched liked paraffins and aromatic compounds [6]. In developing microcrystalline waxes, one may see napthenic (cyclic) and long chained-like paraffins playing a very significant role, and it can be used to determine the growth patterns of macrocrystalline waxes too. The carbon count of paraffin molecules available in the wax deposits has been observed to be higher than 15 atoms [7]. The crystalline structure is formed usually due to the precipitation of wax. These crystals seem thermo-plastic in shape and nature and could be based in solid or liquid depending over the temperature and pressure conditions of crude oil. Modern analytical techniques and tools of today are able to detect up to roughly over 160 carbon atoms in such types of depositions [4, 8].

The most frequent method used for measuring WAT of oil is cross-polar microscopy (CPM) [3], aided by computation using the high-pressure cross-polar microscopy (HPCM) cell apparatus.

In this procedure, the analyte oil is enclosed around a network within the apparatus, until a gelatinous substance begins to develop just below the WAT point. This gel-like formation is known as waxy oil [9]. The wax inhibitor may somewhat modify the procedure and either lower down the interaction tendencies between one crystal to another, or with 3D wax crystal formations known as crystal network formation (CNF). These inhibitors of wax are the most significant in the oil stream industry. The formation of wax can be stopped or prohibited by several techniques including mechanical approach, chemical inhibitors, watering and hot oiling approach, heat reactions, surface coating, microbial procedure, continuous cold-oil flow and magnet-based conditioning of fluid [10].

Deposition of paraffin may be prevented or greatly retarded by the use of chemical surfactants known as 'dispersants'. The choice of batch or continuous treatment depends on the type and number of wells requiring treatment. Woo et al. [11] discussed the use of a mixture of crystal modifiers in a controlled-release matrix and a commercially available ethylene-vinyl acetate copolymer solid. Crystal modifiers and paraffin dispersants are fairly common among the chemicals used, as paraffin crystals alter the crystallization behaviour of waxes. To a large degree, the efficacy of the crystal modifier depends on the type of aggregate-forming waxes present in the liquid. Therefore, selecting a crystal additive is crucial for the effectiveness of the treatment, as they work well in crude oil that is free of water or has low water content. Such polymers co-crystallize with paraffin crystals; in which it will be held in the crude oil in a dispersed position. In this way, the creation of a three-dimensional network occurs at lower temperatures, resulting in a major reduction of the pour point and viscosity of the crude oil.

Heating is a common method used to promote heavy oil transportation through the pipelines. With increasing temperature, the oily content viscosity decreases very rapidly. This is achieved by electrically heating the tube, and isolating the temperature. Heating up the oil to a point where it has a substantial reduction in viscosity is necessary. Technological upgrades such as special design, insulation and welding are necessary to handle this type of modified pipeline transport [12, 13].

Dilution is one of the oldest and most effective ways of reducing oil viscosity and helping in its mobility via pipelines. Condensates, light hydrocarbon naphtha and some organic solvents are the classical diluents used for this method [14]. For this case, the resulting viscosity of the mixture depends on the dilution rate, initial viscosities, and densities of both the diluent and crude oil [15].

Khani et al. categorized four groups of polymeric chemical identifiers used in paraffin, specifically for wax control. The polymers included ethylene-vinyl acetate copolymers (EVAs), polymers of alkylacrylate grafted with EVAs, polyalkylacrylates and derivatives of α-olefin-maleic anhydride copolymers. The use of conventional methods in flow assurance did not prove to be cost-effective. Therefore, research on recent technologies on using other chemical inhibitors is being carried out. These compounds usually improve the flow of crude oil by reducing its pour point. The efficiency of the controlling compound depends on its action which in turn is a function of crude oil and the compound properties [16].

The use of ionic liquids discussed in this chapter is more angled towards the procedures of chemical solvent use. The flow assurance issue is still not studied in-depth for exploring further applications. Thus, the scope was targeted towards ionic liquids in the specific problem of flow assurance. It was seen that ammonium hydroxide dependent ionic liquids and imidazolium-based liquids can mitigate flow assurance constraint by this technique as shown in [9, 17]. The only hurdle left is to achieve actual feasible levels of ionic liquids usage. It is believed that the oil industry should pursue further development of ionic liquid usage, as these compounds can be used in various ways such as wax deposition, asphaltenes precipitation control and in gas hydrate inhibition. Of course, the exact choice of ionic liquid to be used should still be based on the respective chemical properties.

Furthermore, the deposition of wax inside the oil pipelines is one of the biggest issues for transportation of resources from one site to another. It is observed that more than 85% oil production operation lines are affected due to this deposition inside the pipelines [18–20]. The wax that is inside the reservoir's pipelines can block the pores and become one of the top factors in bringing sudden downfall to a supposedly stable production.

Wax deposition can be further sub-categorized into macro- and microcrystalline types [17, 21]. Several procedures and technical methods have been proposed so far for eliminating the obstruction caused by wax deposition in an economically viable manner. There are 20 or more carbon atoms in macrocrystalline waxes, each with difference structures such as a rod shape, plated shape, or even something akin to a thin needle. Whereas in the microcrystalline, one may find the iso-paraffin and branched-chains types of paraffin, along with naphthalene with small contour structures [13].

During the study of this flow issue, scale formation is also considered as one of the factors. In flow assurance, high water cut production is a persistent phenomenon within oil streams. Similar to other ordinary minerals, mineral scale precipitation and scaling is directly coupled with this high water cut production problem. Thus, this phenomenon should be a matter of concern in the industry to ensure steady well production, and to allow for frequent adjustments due to in supply and demand interventions and tube cleaning maintenance [22, 23].

In the oil industry, most regular categories of mineral scales are found such as carbonates and sulphates of barium, calcium and strontium that are in the higher part of the topside facilities. Disrupting the regular transportation system, the growth of depositions by water led to the development of salt or oxide crystalline which are themselves insoluble within the water component of the stream. Moreover, the conventional method where applying scale inhibitors for the prevention of several formations of inorganic salts depends on the clearance of nucleation and growth of salt crystal with the provided procedure [24, 25]. In the literature, readers may find several existing techniques to eliminate scales by either mechanical means, chemical dissolution or scale inhibitors as discussed in section given below.

4.1.2   Scale Deposition Control in Flow Assurance

Hasson et al. tested organic and green antiscalants for scale control. These compounds include polyaminoamide dendrimers, carboxymethylinulin biopolymer (CMI), polyethyleneimine (PEI) and polyethyloxazoline. Some other organic chemicals were also employed such as starch-based polymers, hydroxyethylidene diphosphonic acid in blend with citric acid, isothiazolone and copolymer of acrylate. The inhibition showed by these green scale inhibitors was compared with the commercially used scale inhibitor and it showed competitive results, indicating their effectiveness. Wa et al. used oil-soluble scale inhibitors in water-sensitive wells, and it was noted that suitable hydrocarbon solvents were required for these compounds. MacAdam and Parsons investigated the effect of zinc, copper and iron dosing on $CaCO_3$ nucleation and precipitation. The best inhibitory effect on precipitation of calcium carbonate was obtained by dosing 0.5 mg/L of zinc under the conditions used in the experiment.

As mentioned above, ionic liquids can also be used for scale inhibition through several interrelated modes of action. In contributing towards the scale removal process, the modes of application are not necessarily used only individually; however, they have an effect in amplifying one another. Of such modes for scale control, ones that are identified as important are the following, scale dissolution using IL, heating over IL generation, using IL as carrier for another scale removal agent or acid generation by reacting IL with H_2O. These measures are essentially carried out for scale removal processes. The primary scale dissolution mode is essentially the application of strongly reactive ILs capable of dissolving the solids by contacting it wherever a buildup is present, for example, nearby a well-bore space. In such an example, removal of the carbonate deposits will allow for alternate flow paths in hydrocarbon flow. This allows for the pumping in of other aqueous fluids (such as fresh water), to come into contact with the ILs to generate more acid and aid in further removal of solids. The heat generated from the contact of IL and the fluid will also melt any scale components within the deposit [26].

The potential of IL is well explained by Lakshmi et al. [26], where 1-butyl-3-methylimidazolium hexafluorophosphate [BMIM][PF$_6$] as scale inhibitors are discussed. The scale inhibition efficiency was reported as around 83%, using 25 wt% methylimidazolium hexafluorophosphate [BMIM][PF$_6$] microspheres at laboratory scale, and the mechanism of inhibition morphological study of that scale can be studied using scanning electronic microscope (SEM) or via energy-dispersive x-ray (EDX) analysis. Moreover, the researchers already recommended the usage of [BMIM][PF$_6$] ionic liquid loaded microsphere in oil wells based on several experiments and stated its potential.

4.1.3 Asphaltene Deposition Control in Flow Assurance

Asphaltenes, which are highly complex molecules with the heaviest weight available in crude oils, when deposited are considered much more hazardous than the ordinary wax deposition [17]. These are basically polar compounds, as shown in Fig. 4.1.

The molecules of this substance cause choking inside the oil pipelines and huge reservoirs' wells, thus proving to be hazardous. A highly viscous substance possessing high boiling points, asphaltenes are known as the 'Cholesterol of Petroleum' because of its high proclivity to associate and aggregate [7].

In terms of chemical composition, asphaltenes comprise of heteroatoms like carbon, nitrogen, hydrogen, sulphide and oxygen. Moreover, it forms aromatic structures (cyclic in shape) [27, 28]. Asphaltenes have different solubility ratios with several solvents. It is known to have strong solubility with selected solvents like benzene and toluene; however, it is insoluble in n-pentane and n-heptane at room temperature. Asphaltene can form precipitates in heavy oil, particularly when the temperature and pressure conditions or composition of phases of the oil are manipulated.

The above-mentioned precipitation development can be observed during several stages of recoveries, i.e. primary, secondary and tertiary, in refinery facilities [28]. The precipitation due to asphaltene occurs in undersaturation conditions, more so in light reservoir fluids than heavy hydrocarbon systems. This has led to several problems including a decline in the production rate, and operational problems such as water-oil emulsion and high viscosity. Asphaltene inhibition only comes behind hydrate inhibition, in terms of the expensiveness of resources allocated towards its treatment. [29]. There is mutual consensus in literature and practice that prevention is the best way to solve the problems caused by the deposition of such molecules

Fig. 4.1 Asphaltene lump formation

[30]. The areas in which asphaltene agglomeration in heavy oils is likely to occur are known as asphaltene deposition enveloped (ADE) region or space.

Several researchers tried to identify the asphaltene deposition enveloped region for several crude oil samples, and it was found that the precipitation of asphaltene either due to gas injection or pressure reduction can be quantitatively estimated [25, 31]. The chunk of asphaltene shown in Fig. 4.1 is formed within the pipeline and separated via scale deposition techniques. Madhi et al. [32] studied the effect of three industrial asphaltene inhibitors, namely; cetyl terimethyl ammonium bromide (CTAB), sodium dodecyl sulphate (SDS), and Triton X-100 on two samples of Iranian crude oil. Apart from these chemicals, they also employed non-commercial solutions such as benzene, salicylic acid, benzoic acid, and naphthalene as asphaltene controlling compounds.

4.2 Wax Deposition Control Using ILs Applications

Ionic liquids are conventionally defined under the classes of ionic species of salts that are in the state of liquid at temperatures of less than 100 °C [33, 34]. The significant difference between molten salts and ionic liquids is in the melting point or temperature. The key observable property of ILs is their low melting point.

Other properties of ILs such as high chemical stability in low pressure conditions along with favourable solubility for inorganic and organic compounds play an important role in their applications in different scientific fields. The property of modulated polarity and wide chemical functional group diversity of ILs make them truly unique as well [35, 36]. It is due to this that ILs are investigated for its uses in various domains, including storage, catalysis, biotechnology, petroleum sciences and CO_2 capture [36, 37]. Within the reviewed literature, it is seen that very limited studies are available regarding depositions of wax and its control using ionic liquids. With this in mind, the objective of this chapter is to discuss currently available data on ionic liquids in oil and gas industry applications for wax deposition control. It is understood so far that ionic liquids and its applications can serve as an effective method in removing wax deposition in reservoirs.

4.2.1 Impact of ILs Over Viscosity of Model Oils

Zhao et al. [9] discussed the modification techniques for crude oil via the addition of ionic liquids such as $[BMIM]^+[BF_4]^-$, $[BMIM]^+[PF_6]^-$, $[OMIM]^+[BF_4]^-$, and $[BMIM]^+[Cl]^-$ with crude oil. By adding these ILs, reduction in apparent viscosity is noted but the outcome was not up to expectation. One of the ILs used is D-IL-202, which is commonly used for the same vital role in viscosity reduction. As part of a comparison study by the authors, D-IL-202 has proven to show better performance for wax separation as compared to conventional ILs. The study revealed that the modification within ionic liquids presents greater separation ability for wax, and recommended its use in redesigning mixtures of ionic liquid [9].

Within the same study, a dodecane labelled as 'Model Oil A' is considered to have 5 wt% macrocrystalline. Some of the gel breakage experiments in samples of Model Oil A were performed using 1 wt% $[BMIM]^+[BF_4]^-$ in Model Oil A, 1 wt% $[BMIM]^+[PF_6]^-$ in Model Oil A, 1 wt% $[OMIM]^+[BF_4]^-$ in Model Oil A, and 1 wt% $[BMIM]^+[Cl]^-$ in Model Oil A, which were accomplished at a shear rate of 0.1 s^{-1}. All the practical tests above had been repeated thrice. It was taken into consideration that these wax crystals precipitate at the temperature reading below than WAT, hence the charges of $[BMIM]^+[BF_4]^-$, $[OMIM]^+[BF_4]^-$ and $[BMIM]^+[PF_6]^-$ may stabilize the wax crystals in order to obtain charge-based stable wax colloidal system. The findings above can be explored extensively in the oil industry for its potential in partitioning wax from the crude oil. It is also studied that $[BMIM]^+[BF_4]^-$, $[OMIM]^+[BF_4]^-$ and $[BMIM]^+[PF_6]^-$ influenced the wax crystal network formation in various ways differently. This is because of the polarity and due to the carbon chain length (CCL) of ionic liquids which created an impact on the entire wax crystal formation process. It has been also seen that the $[BMIM]^+[Cl]^-$ influenced the morphology of wax crystals more substantially as compared to the ILs of $[BMIM]^+[BF_4]^-$, $[OMIM]^+[BF_4]^-$ and $[BMIM]^+[PF_6]^-$ [9].

With its presence, the melting point of $[BMIM]^+[Cl]^-$ is shown to be greater than the WAT of model oil samples during the gel process for the model oil A, as the precipitation of crystals serves as nuclei to seed further increase in nucleation and growth. Furthermore, the charges on $[BMIM]^+[Cl]^-$ may transform a charge based stable colloid system. In the entire gelation process, very ardent and stable crystals network can be formed and observed. The properties of $[BMIM]^+[Cl]^-$ as compared to other ILs are much better to be used as wax promoters, and may be considered for wax separation purposes in the oil industry. Figure 4.2 presents the performance of $[BMIM]^+[Cl]^-$, demonstrating its superior results compared to other ionic liquids.

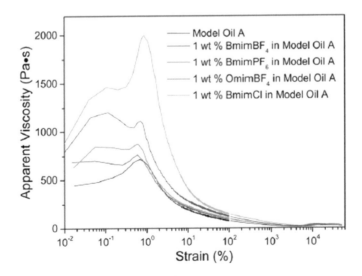

Fig. 4.2 Graph of apparent viscosity at 0.1 s^{-1} rate vs Model oil A [9]

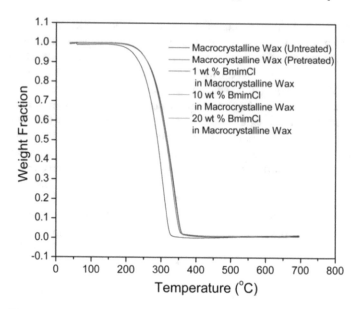

Fig. 4.3 Thermogravimetric curves of untreated vs pre-treated macrocrystalline wax [9]

4.2.2 Impact of Wax State Over WDT

From Zhao et al.'s work, the influence of the initial wax state conditions on the wax decomposition temperature was investigated, with all the experiments are performed in duplicate. Two protocols were used; in the first, the untreated wax is added to the differential scanning calorimetry (DSC) pan directly, while the second protocol heated the wax to liquid at 110 °C for 30 min prior to further treatment.

Figure 4.3 shows the waxes' profile, where the WDT is nearly 259 °C for the untreated wax, while the WDT for the heat-treated wax around 279 °C.

4.3 ILs as Asphaltene Deposition Control Compounds

Few techniques are readily available in removing asphaltenes from any sort of reservoir or pipelines. The main issue is in the different insoluble ratios, e.g. the ILs are insoluble in lower alkanes such as n-pentane and n-heptane, but are workably soluble in aromatic solvents like toluene, benzene, or xylene [10]. But it has been understood that the problem of asphaltene precipitation or deposition in pipelines is a significant problem being faced today by the industry [21].

If left alone, consequences include an overall shutdown of the well, and ultimately major losses in production, alongside the financial burden of remediation and cleaning cycles [27]. Moreover, these depositions are not easy to tackle and can be harmful as they are not environmentally friendly [38], thus the industry turned to surfactants [38]. However even within the literature of surfactant studies, limitations still exist. Surfactants are unable to dissolve asphaltenes containing higher molecular weights, hence the attention is shifted towards ILs, which not only can dissolve the targeted compounds, but also have the capability to disperse effectively in crude oils [39, 40]. Moreover, these ILs are dispersants in nature. The asphaltene deposition by ILs interacts through charge transfer, rather than interactions between neighbouring aromatic rings (pi-pi interactions).

It is also noted that acyclic hydrophenancerence ring will react more actively with asphaltenes than the other simple alkyls and alkyl phenols present in ILs [17]. It is also noted that application of ionic liquid for the dissolution of this heavy molecule is achievable in both reservoirs and pipelines as well.

In order to investigate this mechanism, Boukherissa et al. [39] studied (1-propyl boronic acid-3-alkylimidazolium bromides and 1-propenyl-3-alkylimidazolium bromides) as dispersants of petroleum asphaltenes. Moiety of the boronic acid, and type of alkyl chain present in the IL, and length of side chains present in the IL play a role in controlling the deposition and aggregation of the asphaltenes. In another study, Zheng et al. [41] proposed the treatment of asphaltenes with 16 different ionic liquids that had three major cationic charges such as imidazolium, cholinium and phosphonium with a variety of anions. In the elemental analysis, through thermogravimetric analysis (TGA) and Fourier-transform infrared spectroscope (FTIR) spectra of the asphaltenes, it was seen that the aliphatic and oxygen content of the recovered asphaltenes (after it was treated with the ILs) was higher than the raw asphaltene samples, indicating that the ILs was able to capture the dissolved/suspended asphaltenes, even for small particles that were not captured by the 0.2 μm filter in place. Therefore, the attractive property of the ILs being fully oil miscible makes it a viable alternative compared to the more toxic surfactants such as toluene or xylene.

In case of asphaltene precipitation control, the synthesis of imidazolium-based ionic liquids like 1-buytl-3-methylimidazolium chloride, 1-butyl-3-methylimidazolium nitrate, and 1-methyl-1H-imidazol-3-ium-2-carboxybenzoate has been tested and the 1-buytl-3-methylimidazolium chloride showed the highest asphaltene dispersion performance compared to the other compounds [41]. These studies showed that the field of ionic liquids application in asphaltene control shows a promising future, and more suitable compounds can be further tested and explored. Zhao et al. [9] investigated the impact of [BMIM][Cl] on asphaltene decomposition temperature. The investigation entails the use of TGA for a temperature parameter range between 35 and 700 °C, at a specific heating rate of 10 °C/min. On a tangential note, the study was noted to be able to reproduce the experimental results very well. Figure 4.4 shows the decomposition temperature of [BMIM][Cl] and asphaltene molecule, which is at 270 and 423 °C approximately. It was concluded that for the mixture of 45.90 wt% [BMIM][Cl] in asphaltene compound, the decomposition temperature reading or value is 225°C, which is lesser than that of pure [BMIM]$^+$[Cl]$^-$.

One of the initial findings included the influences of ILs and amphiphiles on the mitigation of heavy complex molecules within asphaltene precipitation, in oil reservoirs with the presence of CO_2. In this study, oil and CO_2 miscibility was

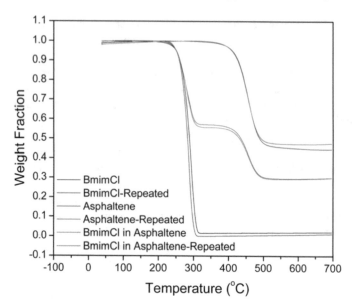

Fig. 4.4 Thermal gravimetric curves of Proposed ILs, asphaltene, vs 45.9 wt% $[BMIM]^+[Cl]^-$ in asphaltene [9]

observed. ILs with different kinds of cationic groups and anions were and with a focus on the effect of cation chains, group heads and the amalgamation of cations and anion. The impact of ionic liquids in terms of concentrations had been also studied [41].

After performing the studies on oil with CO_2 in miscible conditions, pressure change and precipitation were also measured quantitatively. This study showed that ILs containing p-alkylpyridinium cation with a chloride anion performed better, and a decrease in performance was observed with decreasing alkyl chain length. The inhibition was observed according to the following order: $[C_{12}py]$ $[Cl]$ < $[C_8py]$ $[Cl]$ < $[C_4py]$ $[Cl]$. Shorter chain cations showed a better inhibition performance than longer chain length cations. While checking the similarities and dissimilarities of ionic liquids of different cations but the same anion, $[C_4iql]^-$ $[Cl]^-$ showed better effectiveness than $[C_4py]$ $[Cl]$ in inhibiting CO_2 instigated asphaltene precipitation.

Liquid fuels can come directly from coal, and one of such conversion processes is called direct coal liquefaction (DCL). The asphaltenes that make up 25 wt% of the DCL products are loaded with unwanted aromatic precursors, which needs to be extracted from the DCL products prior to shipment. Bai et al. studied a series of protic ionic liquids for this exact extraction requirement, using N-methylimidazole, 3-methylpyrdine, and trimethylamine cations [42]. The acetate, benzoate anions and propionate anions used to extract these asphaltenes from DCL items were conducted at room temperature, while also providing higher yields over ordinary solvents. The ratio of the extracted content consists of lower hydrocarbon content and with higher degrees of aromatics, but lesser contents of sulphur, no ash contents, and essentially no presence of the unwanted quinoline insolubles. Matching with what was discussed before, with increasing chain length of anions, the extraction yields become even greater. However, the strength of the cations used in the protic ILs as per the following order of: N-methylimidazolium[MIM]$^+$<triethylammonium[TEtA]$^+$ < 3-methylpyridinium [MPy]$^+$. The mechanisms that took assisted in dissolving the asphaltenes were reportedly a mix of Π-cation interactions, hydrogen bonding, and transfer of charges between complexes.

IRAN91, which is an ionic liquid compound formed by mixing [Et$_3$N] HCl with AlCl$_3$, was studied by Joonaki et al. for heavy oil recovery through removal of asphaltenes [43]. This compound is observed to be most effective in decreasing the asphaltene content, as observed by both the reduction in average molecular weight and viscosity of the heavy oil. The viscosity observed to be decreased from 1800 to 645 cP, whereas the weight went down from 2840 to 384 gr.(gr.mol)$^{-1}$. Lastly, the total asphaltene present in the treated oil was reduced to 7%, from the original amount of 15% in the raw heavy oil. This occurs through the formation of a complex between IRAN91 and the sulphur in the oil, weakening the bonds between the carbon and sulphur atoms. Thus, as the asphaltene count decreases in the oil, leading to a fractional increase of the saturates and aromatics within the treated oil, prompting the reduction in viscosity for the bulk oil, thus initiating the changes within the oil's conditions as mentioned. Therefore, due to the strong attraction between the cationic components of the IL towards the sulphur within the oil, the authors concluded that a higher sulphur count in the oil can be helpful in further improving its quality during the IL treatment.

Ogunlaja et al. used imidazolium cations with different combination of anions and studied how the different combinations influenced its capabilities in dispersion of asphaltenes. The ILs synthesized and later utilized in the study included 1-butyl-3-methylimidazolium chloride, 1-butyl-3-methylimidazolium nitrate, and 1-methyl-1H-imidazol-3-ium-2-carboxybenzoate [44]. Through numerical analysis and simulations, the complex reactions between the asphaltenes and the ionic liquids were found similarly enough to agree with the work of Bai et al. [42].

On the basis of these reactions, researchers found the energies of 1-butyl-3-methylimidazolium chloride, 1-butyl-3-methylimidazolium nitrate and 1-methyl-1Himidazol-3-ium-2-carboxybenzoate to be -55.4×10^4, -44.1×10^4 and -54.8×10^4 kcal/mol, respectively. It was also observed from the interaction energies that the smallest reaction energy of 1-butyl-3-methylimidazolium chloride was comparatively better at dispersing asphaltenes.

One may look over the formula of the heavy oil before and after the reaction epoch with the proposed ILs as mentioned in Table 4.1. The saturates, aromatic, resins and asphaltenes content within the heavy oil prior to treatment are found to be 28.4, 27.6, 32.4 and 11.6%, respectively.

After the reaction, all contents except for the asphaltenes showed an increase in terms of constituent. The reaction is illustrated as below:

Table 4.1 Impact of ILs over the composition of asphaltene [44]

System	Asphaltene composition (wt%)
Untreated oil	11.6
[(Et)$_3$NH][AlCl$_4$]	9.5
[(Et)$_3$NH][AlCl$_4$]$^-$Ni^{2+}	6.6
[(Et)$_3$NH][AlCl$_4$]$^-$Fe^{2+}	29.2
[(Et)$_3$NH][AlCl$_4$]$^-$Cu$^+$	37.3

$$\text{Resin} \xrightarrow[\text{ILs}]{} \text{light hydrocarbons} + \text{asphaltene}$$

It can be observed from Table 4.1 that the metal modified ionic liquids have an impact over the upgradation of the heavy oil sample. From the above-mentioned discussion, it is seen that the ILs possess potential in keeping the asphaltene within the provided solution.

Rashid et al. studied the potential effect of ionic liquids on asphaltene control using COSMO-RS. It was argued that due to the complex nature of asphaltene, extraction using ordinarily used volatile organic compounds (VOCs) remained unproductive, however the use of ILs as solvents for its removal from various fluid streams are being investigated and proved to be a promising substitute. The most significant factors regarding the ionic liquids and their performance parameters in the respective field comprise the steric hindrance effect, hydrogen bonding, and polarity of cations and anions [45].

4.4 Summary

It is evident from the literature review that the use of ILs in wax inhibition has been studied and has a promising prospect. Different methods have previously been employed to tackle the issue, which include mechanical and chemical approaches. The chemical method is preferred due to being non-disruptive. In one study, ILs were used as wax promoters, to ensure that a swift separation can be performed afterwards. Imidazolium-based ILs were used with different combination of cations including chloride, tetrafluoroborate and hexafluorophosphate cations. Concerning the use of ILs, it has been found and accepted that the acid moiety is essential for the stabilization of asphaltenes. In addition to this, it is quite clear the inhibitors of asphaltene are complex and rely on various elements. The overall efficiency percentage relies on not only the chemical characteristics of asphaltenes but also on the nature of the solvent, i.e. oil, particularly in terms of its quality. Thus, it is argued that a generalized inhibitor that is made to be efficient with most types of crude oil samples is not realizable. The chemical properties of the inhibitors should be considered prior to be used for any particular crude oil sample. Ionic liquids are well found to be suited for the use in wax deposition, primarily due to two main factors. First, they are easily soluble in the crude oil, and secondly the molecules of the ILs are capable of forming charge transfer complexes and hydrogen bonding by opting required amount of cations and anions in lateral chains.

The reaction between asphaltenes and ILs can result in different geometries and complexes, hence it is necessary to derive a relationship between the parameters and the inhibition of ionic liquids. This manuscript demonstrates that these types of ionic liquids can be used to control the precipitation occurred by asphaltene aggregates. As demonstrated by ILs containing boronic acid moiety, which replaced the lateral chain of ILs to bond with the asphaltene contaminants, thus generating an important reduction of size. Moreover, the smallest aggregates with sizes of 20 nm were obtained with ILC16. Thus, the class of dispersants was found to act in many different ways; some act directly over the molecular level, whereas other less active and less efficient dispersants such as alkylphenols adsorb at the superficial level, with the big aggregates forming solvation shells.

Acknowledgements The authors would like to acknowledge Chemical engineering department and CO2RES center, Universiti Teknologi PETRONAS, Malaysia for their support.

References

1. Zhu T, Walker JA, Liang J, Laboratory PD (2008) Evaluation of wax deposition and its control during production of Alaska North Slope oils, no. December
2. Wilson A, Overaa SJ, Holm H (2004) Ormen lange - Flow assurance challenges. Proc Annu Offshore Technol Conf 2:905–914
3. Wei B, Lu L, Li H, Xue Y (2016) Novel wax-crystal modifier based on β-cyclodextrin: synthesis, characterization and behaviour in a highly waxy oil. J Ind Eng Chem 43:86–92
4. Pedersen KS, Rønningsen HP (2003) Influence of wax inhibitors on wax appearance temperature, pour point, and viscosity of waxy crude oils. Energy Fuels 17(2):321–328
5. Cardoso CB, Alves IN, Ribeiro GS (2003) Management of flow assurance constraints. Proc Annu Offshore Technol Conf 2003–May, pp 1430–1438
6. Singh A, Lee HS, Singh P, Sarica C (2011) Flow assurance: validation of wax deposition models using field data from a subsea pipeline
7. Theyab MA (2018) Fluid flow assurance issues: literature review. SciFed J Pet 2(1):1–11
8. Edmonds B, Moorwood T, Szczepanski R, Zhang X (2008) Simulating wax deposition in pipelines for flow assurance. Energy Fuels 22(2):729–741
9. Zhao Y, Paso K, Zhang X, Sjöblom J (2014) Utilizing ionic liquids as additives for oil property modulation. RSC Adv. 4(13):6463–6470
10. Velusamy S, Sakthivel S, Gardas RL, Sangwai JS (2015) Substantial enhancement of heavy crude oil dissolution in low waxy crude oil in the presence of ionic liquid. Ind Eng Chem Res 54(33):7999–8009
11. Woo GT, Garbis SJ, Gray TC (1984) Long-term control of Paraffin deposition. Proc - SPE Annu Tech Conf Exhib 1984–Sep, no. 1
12. Dauphin C, Daridon JL, Coutinho J, Baylère P, Potin-Gautier M (1999) Wax content measurements in partially frozen paraffinic systems. Fluid Phase Equilib 161(1):135–151
13. Hammami A, Ratulowski J (2007) Precipitation and deposition of asphaltenes in production systems: a flow assurance overview. Asph Heavy Oils, Pet, pp 617–660
14. Bassane JFP et al (2016) Study of the effect of temperature and gas condensate addition on the viscosity of heavy oils. J Pet Sci Eng 142:163–169

15. Gateau P, Hénaut I, Barré L, Argillier JF (2004) Heavy oil dilution. Oil Gas Sci Technol 59(5):503–509
16. Mahmoudkhani A, Feustel M, Reimann W, Krull M (2017) Wax and paraffin control by fracturing fluids: understanding mode of actions and benefits of water-dispersible wax inhibitors. Proc - SPE Int Symp Oilf Chem 2017–April, pp 995–1012
17. Bera A, Agarwal J, Shah M, Shah S, Vij RK (2020) Recent advances in ionic liquids as alternative to surfactants/chemicals for application in upstream oil industry. J Ind Eng Chem 82:17–30
18. Chua PC, Kelland MA (2012) Tetra(iso-hexyl)ammonium bromide - The most powerful quaternary ammonium-based tetrahydrofuran crystal growth inhibitor and synergist with polyvinylcaprolactam kinetic gas hydrate inhibitor. Energy Fuels 26(2):1160–1168
19. Kelland MA (2006) History of the development of low dosage hydrate inhibitors. Energy Fuels 20(3):825–847
20. Sloan ED (2005) A changing hydrate paradigm - From apprehension to avoidance to risk management. Fluid Phase Equilib 228–229:67–74
21. Elganidi I, Elarbe B, Ridzuan N, Abdullah N (2020) A review on wax deposition issue and its impact on the operational factors in the crude oil pipeline, vol XX, no X
22. Fink J (2003) Oil field chemicals, 1st edn. Elsevier
23. Huang HJ, Ramaswamy S, Tschirner UW, Ramarao BV (2008) A review of separation technologies in current and future biorefineries. Sep Purif Technol 62(1):1–21
24. Shaban S, Dessouky S, Badawi AEF, El Sabagh A, Zahran A, Mousa M (2014) Upgrading and viscosity reduction of heavy oil by catalytic ionic liquid. Energy Fuels 28(10):6545–6553
25. fu Fan H, bao LI Z, Liang T (2007) Experimental study on using ionic liquids to upgrade heavy oil. Ranliao Huaxue Xuebao/J Fuel Chem Technol 35(1):32–35
26. Lakshmi DS, Senthilmurugan B, Drioli E, Figoli A (2013) Application of ionic liquid polymeric microsphere in oil field scale control process. J Pet Sci Eng 112:69–77
27. Turosung SN, Ghosh B (2017) Application of ionic liquids in the upstream oil Industry-A review. Int J Petrochemistry Res 1(1):50–60
28. Bahman J, Sharifi K, Nasiri M, Haghighi Asl M (2018) Development of a log-log scaling law approach for prediction of asphaltene precipitation from crude oil by n-alkane titration. J Pet Sci Eng 160:393–400
29. Brown LD (2002) Flow assurance: A π 3 discipline. Proc Annu Offshore Technol Conf, pp 183–189
30. Peng X, Hu Y, Liu Y, Jin C, Lin H (2010) Separation of ionic liquids from dilute aqueous solutions using the method based on CO_2 hydrates. J Nat Gas Chem 19(1):81–85
31. Gao S (2008) Investigation of interactions between gas hydrates and several other flow assurance elements. Energy Fuels 22(5):3150–3153
32. Madhi M, Kharrat R, Hamoule T (2018) Screening of inhibitors for remediation of asphaltene deposits: experimental and modeling study. Petroleum 4(2):168–177
33. Welton T (1999) Room-temperature ionic liquids. solvents for synthesis and catalysis. Chem Rev 99(8):2071–2083
34. Zare M, Haghtalab A, Ahmadi AN, Nazari K, Mehdizadeh A (2015) Effect of imidazolium based ionic liquids and ethylene glycol monoethyl ether solutions on the kinetic of methane hydrate formation. J Mol Liq 204(04):236–242
35. Smiglak M et al (2014) Ionic liquids for energy, materials, and medicine. Chem Commun 50(66)
36. Ratti R (2014) Ionic liquids: synthesis and applications in catalysis. Adv Chem 2014(3):1–16
37. Nashed O, Partoon B, Lal B, Sabil KM, Mohd A (2018) Review the impact of nanoparticles on the thermodynamics and kinetics of gas hydrate formation. J Nat Gas Sci Eng 55:452–465
38. Mullins OC (2011) The asphaltenes. Annu Rev Anal Chem 4(1):393–418
39. Boukherissa M, Mutelet F, Modarressi A, Dicko A, Dafri D, Rogalski M (2009) Ionic liquids as dispersants of petroleum asphaltenes. Energy Fuels 23(5):2557–2564
40. Atta AM, Ezzat AO, Abdullah MM, Hashem AI (2017) Effect of different families of hydrophobic anions of imadazolium ionic liquids on asphaltene dispersants in heavy crude oil. Energy Fuels 31(8):8045–8053

41. Zheng C, Brunner M, Li H, Zhang D, Atkin R (2018) Dissolution and suspension of asphaltenes with ionic liquids. Fuel 238:129–138

42. Bai L, Nie Y, Li Y, Dong H, Zhang X (2013) Protic ionic liquids extract asphaltenes from direct coal liquefaction residue at room temperature. Fuel Process Technol 108:94–100

43. Joonaki E, Ghanaatian S, Zargar G (2012) A new approach to simultaneously enhancing heavy oil recovery and hindering asphaltene precipitation. Iran J Oil Gas Sci Technol 1(1):37–42

44. Ogunlaja AS, Hosten E, Tshentu ZR (2014) Dispersion of asphaltenes in petroleum with ionic liquids: Evaluation of molecular interactions in the binary mixture. Ind Eng Chem Res 53(48):18390–18401

45. Rashid Z, Wilfred CD, Gnanasundaram N, Arunagiri A, Murugesan T (2018) Screening of ionic liquids as green oilfield solvents for the potential removal of asphaltene from simulated oil: COSMO-RS model approach. J Mol Liq 255:492–503

Printed in the United States
By Bookmasters